fashioning

ics

fashioning
fabrics

contemporary
textiles
in fashion

Black Dog Publishing

contents

the hidden art of fashion

sandy**black**

There are many texts on fashion designers, and many texts on textiles, but few which integrate fabrics with fashion. *Fashioning Fabrics Contemporary Textiles in Fashion* celebrates the place of textiles in fashion, and starts to redress this imbalance by foregrounding the materiality of fashion. Both established designers and emerging talents are featured, working across a wide range of techniques and processes for constructing, forming, sculpting, embellishing, patterning, treating and mis-treating textiles.

Although it seems obvious that textiles are an intrinsic foundation of clothing, the invention, development and manipulation of fabrics has long remained the hidden art of fashion. Fashion designers choose fabric for its many and varied qualities of touch, handle, drape and aesthetics, (within the price ranges open to them), and are extremely sensitive to the fabric's behaviour or potential to achieve a desired silhouette. Tacit knowledge is built up over a period of time, enabling the designer to visualise (and sometimes work against) the inherent characteristics of the textiles they have selected, which often may not be seen in volume until first prototypes. On another level those who are 'createurs' in the couture tradition will personally mould, drape and sculpt a fabric around the body to create a form, and have the means to commission wonderful and exclusive fabric from the most esteemed textile houses. In the French couture ateliers armies of the famous '*petites mains*' —skilled embroiderers and seamstresses— manufacture the designer's concepts painstakingly by hand. These are the flights of fancy gracing the most select catwalks, the extravagant visions seen today in the collections of Dior, Lacroix, Galliano, Gaultier, Balenciaga, Givenchy, and Chanel for example, creating and maintaining the image upon which the fragile marketing edifice of high fashion is built, sustained by sales of perfume, accessories and other aspirational items. The original couturiers of Paris, London and Florence (the first fashion city of Italy, before Milan, where shows took place in the beautiful Renaissance Pitti Palace) operated within a network of highly specialised textile producers such as weavers Bianchini-Ferier and Boussac, printers Ascher, and embroiderers Lesage, creating sumptuous laces, beadwork and voluptuous fabrics of all kinds. Today this relationship has continued with fabric specialists such as Jacob Schlaepfer of Switzerland, and Zibetti, Ratti and Etro in Italy, who create unique woven and embellished textiles for design houses.

But fashion is now a broad church, operating on a spectrum from couture and designer brand level, utilising the highest quality materials, to the mass market where pricing is key and fabrics are sourced on cost as much as

materials and processes of fabric development to break new ground and influence the direction fashion is taking. The creation of a clothing collection is a highly collaborative activity; many fashion designers work closely with specific textile designers, but these sources are not usually publicly acknowledged, forming the invisible backbone of the industry by feeding inspirational ideas into fashion through the networks of trade fairs and consultancy. Consequently, there exists a symbiotic but not necessarily equal relationship between fashion designers and textile designers.

Newly graduated textile designers are a rich source of inventive concepts, and are often commissioned to produce work for a particular season's catwalk collection, creating a signature look which informs and influences the fashion designer's vision. An indivisible bond is thus forged between the fabrication and the fashion image and silhouette. Often, long after the clothes have been discarded and the designer has moved on, an iconic fashion image remains, capturing a moment in time. For example, knitwear designer Sidney Bryan graduated from the Royal College of Art in 1997 with a collection of oversized men's knitwear and was immediately commissioned to create showpieces for the Alexander McQueen catwalk, which he continued to do for a number of seasons. Similarly, Alex Gore-Browne produced special versions of her degree collection knitwear for McQueen, hand-sewn with 12,000 felt 'sequins'—an echo of the French atelier system, but peopled instead by fashion and textile students.

Talent scouts from major companies such as Donna Karan in the USA, and Alberta Ferreti in Italy, scour the UK degree shows and the Paris

for any other quality. Recently, the fashion spectrum has both polarised and expanded to include the 'value' market where clothing has become a basic commodity, produced and sold at the highest volume and cheapest price in supermarkets and discount stores. None of this would be possible without the global sourcing and individual labour of textile and garment workers around the world, especially in China and India. However, increasingly, the discerning, knowledgeable and globally aware consumer wishes to know something about the pedigree of items before purchasing—their manufacturing cycle and ecological credentials—and interest in how and where textiles are produced fuels the desire for authenticity and craftsmanship.

Between the extremes of these market levels there is a significant and diverse group of innovative fashion designers and emerging designer-makers who experiment with the

Zibetti, 1992
ruched satin fabric
embroidered using
elastomeric thread
photo: Sion Parkinson
courtesy Sandy Black

fabric trade fairs for upcoming designers, buy fabric design samples, and then attempt to translate the liveliness of the student designs into commercial production. In recognition of the importance of the annual crop of new design talent, Donna Karan mounts a touring show of fresh work in their flagship stores and sponsors prizes and competitions. Some fortunate designers are also offered jobs. Helen Amy Murray, a graduate of Chelsea College of Art and Design, who invented an award-winning cut-out textile process by which she creates impressive relief textures of floral designs in suede and leather, was commissioned by Karan to create a showpiece mannequin.

Several new designers, such as Louise Goldin, Kanako Kajihara, and Sanghee Chun are featured in *Fashioning Fabrics*, having created a body of work with a handwriting that will continue to develop through their personal journeys into professional fashion and textile environments, walking a difficult line between creative individuality and commercial realities. When these aspects are successfully combined—realistic cost together with aesthetic qualities of concept, surface, drape, colouration, proportion—then the work can become truly influential as it permeates the industry.

These graduates are continuing an older tradition of external artist and designer collaborations with the textile industry, which were particularly strong in the early twentieth century. Following the Industrial Revolution, by the end of the nineteenth century, many new schools of design had been set up in Britain with the express intention of providing the industry with commercial artists (i.e. designers) to improve the quality of commercial textiles, and to reconnect technique with aesthetics.[1] European textile manufacturers also commissioned recognised artists such as Sonia Delaunay (who believed fine arts should be integrated into everyday life) and Henri Matisse, or the burgeoning design workshops such as the Bauhaus in Germany, Wiener Werkstatte in Vienna, and Omega Workshop and Silver Studio in London.

Pioneering couturier Paul Poiret set up the Atelier Martine to create decorative arts and fabrics for his collections and for interiors, using designs by artist Raoul Dufy for a time, who had also created textile designs for silk weavers Bianchini-Ferier of Lyons until 1930. Before the First World War, Diaghilev's Ballets Russes was highly influential and much design inspiration came from the folk patterns of Greece, Russia and Egypt (inspired by the discovery of the tomb of Tutankhamen), in addition to the already popular exoticism of

Helen Amy Murray
Donna Karan showpiece
mannequin, 2003

Chinoiserie, and Indonesian batik patterning. In 1875, Arthur Lasenby Liberty had opened his Liberty emporium, which introduced oriental decorative arts including textiles and carpets to London society and was highly fashionable amongst members of the aesthetic movement. Original textiles for fashion and interiors were developed to challenge the dominance of Paris, and continued to be a mainstay of the Liberty style, providing a showcase for many textile designers.

The Bohemian set loved rich fabrics such as velvets and chenilles, and particularly the fluid columns of the extraordinary Delphos dresses by Mariano Fortuny, pleated and dyed in burnished colours, which were exhibited at the Paris Exhibition of 1925, the pinnacle of the Art Deco movement. The secrets of Fortuny's textile process have never been discovered, although Wales-based designers Charles and Patricia Lester produce a very close approximation. However, the contemporary equivalent of the Delphos dress is undoubtedly Issey Miyake's *Pleats Please* range.

During the decade following the Paris Exhibition, socialite and designer Elsa Schiapparelli famously collaborated with the Surrealist artist Salvador Dalí to bring into textiles and clothing some of his imaginative visual ideas, such as the Desk Suit, echoing the painting *Venus de Milo with Drawers* and the *trompe l'oeil* print of the Tear dress, designed to look like the dress was ripped, exposing flesh beneath, a shockingly violent statement at the time. Surrealism has continued to influence the concept, staging and design of fashion, both directly and indirectly; for example, the chiffon-wrapped faces of Martin Margiela's catwalk models of

1995 can be directly referenced to the Magritte painting *The Lovers*. Margiela subverts expectations in his use of textiles and fabrication—his own version of *trompe l'oeil* recreates clothing as simulacra, for example printing the monochrome image of a sequined dress or a knitted pullover on a plain satin fabric dress or top.

Art Deco style was expressed by brightly coloured, playful, and essentially positive designs, reflecting the fundamental optimism of a new Modernist age. Print design motifs were often narrative, directly reflecting aspects of life—as seen in the Constructivist designs of the machine age—a trait maintained particularly in the USA, in the novelty or 'conversational' prints of the pre Second World War period. This concept continued in wartime, with scarves produced by Jacqmar in England carrying propaganda messages such as "Salvage your Rubber".[2]

Jacqmar printed wool scarf, c 1943
photo: Sion Parkinson
courtesy Sandy Black

lace and crochet, both by hand and machine, allow infinite combinations of yarns, experimental materials and structures, using natural, synthetic or combined materials. Non-woven fabrics may also be made from fibres which have been bonded together by hand or industrially, such as felts or glass fibre fabrics; others are made from thin films, which can be shredded to create fibres and yarns.

In the immediate post war period of the 1950s and into the radical 60s, the space age inspired the use of modern futuristic materials in fashion, including the new plastics and synthetic yarns. Revolutionary ideas in materials were shown by the French designers Pierre Cardin, Paco Rabanne and Courreges. Having worked in the Paris couture houses of Paquin and Dior, and also with Elsa Schiapparelli, Cardin opened his own house in 1950. By the mid-60s he had moved from classic wools and tweeds to embrace plastics, PVC and the more rigid qualities of synthetic double jersey and heavyweight wool fabric. His sculptural, geometrically-cut, streamlined designs—which were sometimes heat-moulded to three-dimensional surface effects in a fabric he named Cardine—gave the feeling that the clothes would have stood up by themselves.

Today, the slogan tee-shirt has taken over as the main vehicle for comment, but without the powerful political impact of the anti-war statements made on the catwalk by Katharine Hamnett in 1984. Designers do, however, continue to use print in a subversive way, as seen in the controversial imagery used by designers Basso & Brooke.

The textile surface offers a vast range of possibilities for treatments, whether embellishments such as printing, beading, embroidery or appliqué, special surface finishes such as latex coating, needle punching, or laser etching and cutting, or manipulation by pleating, dyeing (including resist dye and *shibori* or tie-dye) and heat setting. Alternatively, the construction of the fabric from first principles offers even more scope for unique pieces: weaving, knitting,

Paco Rabanne, previously working in costume jewellery, created 'chain mail' clothes from linked plastic and metal lozenges, which were adjusted not by sewing but by manipulating their metal links with pliers. His modular constructions expanded the boundaries of what might be considered textiles, and have continued to influence designers to create showpieces using any conceivable material. For example, in the 1970s and 80s Jean-Charles de Castelbajac's extreme fashions adopted the humour of Pop Art, making a fur coat from

Basso & Brooke
The Garden of Earthly Delights, 2004

made of real banknotes in 2001. Brazilian designer Alexander Herchcovitch showed Rabanne-inspired clothes in 2000, and both mens- and womenswear made from feathery shredded PVC for Autumn/Winter 2001. Recent London College of Fashion graduate Rishti Diwan made a collection entirely from draped metal chains, discs and curved bands, coming full circle to Paco Rabanne, whose original modular dresses have been revived for Autumn/Winter 2005.

assembled teddy bears, or a jacket from soup cans. Other examples include Martin Margiela's waistcoat made of crockery pieces from 1996; Alexander McQueen's razor clamshell dress from the 2000 *Voss* collection; Andrew Groves' 1998 debut collection with razor blades dress and headpiece made of six inch nails; Julien Macdonald's knitted dress festooned with flourescent wands in the 1999 *Metallurgical* collection; and Russell Sage's £6,000 dress

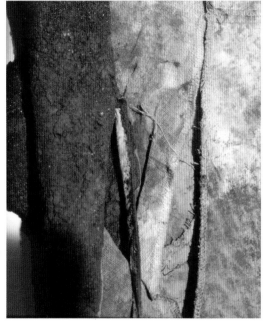

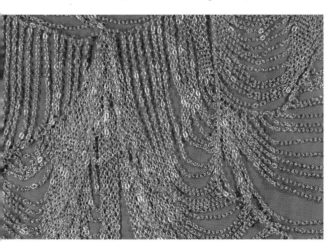

In the last decades, the most avant-garde designers have used increasingly bizarre fabric treatments. For example, Hussein Chalayan's graduation collection is remembered for his experiments with burying fabrics covered with iron filings in the garden; Martin Margiela 'grew' fabrics with bacteria, covering and changing the surface over a period of time, during the Rotterdam exhibition *9/4/1615* in 1997. Influential designers Rei Kawakubo of Comme des Garçons, and Yohji Yamamoto, were the first international designers to

top left
Russel Sage
Autumn/Winter 2001
photo: Anthea Simms

bottom left
Rishti Diwan
London College of Fashion
graduate collection, 2005

right
Hussein Chalayan
detail of buried fabric
Autumn/Winter 1994

'deconstruct' clothing, exposing their inner seams and makings, and using boiled, wrinkled, unfinished and distressed textile treatments in their voluminous clothes. An early hand-knitted 'lace' sweater design from Kawakubo included randomly made holes (neatly cast off and cast on—which after punk could simply have been slashed), giving the impression of being worn out.

Kawakubo's uncompromising design vision has redefined both cloth and silhouette, and continually moves the fashion agenda forward. The Autumn/Winter 2005 collection utilised digital printing on a series of wedding style dresses, designed to look like the folds of another superimposed dress. Protégé and design collaborator Junya Watanabe developed his own line in 1992 under the Comme des Garçons label to express his often extreme, inventive fabrications and radical silhouettes, such as the honeycombed sculptural forms reminiscent of paper toys created in layers of tulle for Autumn/Winter 2000. His treatment of denim as a couture fabric for Spring/Summer 2002, crafted into intricate cuts, was universally applauded. Other collections have featured boiled tartans, padded duvet shapes, and wool tweed bonded to vinyl for reversible and waterproof coats, dresses and skirts.

Martin Margiela picked up the baton of deconstruction to create his own language and taxonomy of clothes, turning them literally inside out by using lining materials and half finished effects. He also constructed garments from recycled items, including a waistcoat made from discarded gloves and sweaters made from army surplus socks.[3] Such textile treatments were shocking at first, but have

Comme des Garçons
Autumn/Winter 2005
photo: Anthea Simms

Junky Styling
Mac Apron Dress, 1998
dress made from a
macintosh raincoat
photo: Luz Martin

12

now become commonplace: ripped jeans, raw, unfinished and frayed edges, and crumpled clothes are considered normal in the ranges of both designer and mainstream fashion on the high street.[4] Recycled and re-worked 'vintage' (the new term for second hand) clothing has also entered mainstream fashion retail stores, with an increasing number of small-scale UK-based designers working in this area, such as Jessica Ogden and Junky Styling.

The arrival of the Japanese designers Issey Miyake, Comme des Garçons and Yohji Yamamoto to the Paris schedules at the start of the 1980s caused a sensation and began to redefine Western views of fashion. Earlier *émigré* designers Kansai Yamamoto and Kenzo Takada had quietly established their businesses from a Paris base, but in contrast this triumvirate of 'trans-national' designers already had businesses in Japan, and continue to work between Paris and Tokyo, producing fabrics in Japan, India and Europe. Both Kansai and Kenzo, whose directional label Jungle Jap was established in 1970, were noted for colourful and multi-patterned collections, using floral and other decorative printed motifs, mixing an identifiable Japanese iconography with a European sensibility and inventive designs. Kenzo is particularly known for intensely coloured and patterned scarves, folk embroideries and multi-colour, intricately knitted intarsias and jacquards, taking inspiration from cultures all over the world, from Russia to South America.

Issey Miyake took a different, less controversial route than his contemporaries, preferring not to be termed a fashion designer. Having experienced both the French couture industry and American commercial fashion, he concentrated on experimenting in simplified

forms—such as 'one piece of cloth' textile construction—and with treatments and materials, together with Makiko Minagawa, his chief textile designer since the inception of the Miyake Design Studio in 1970. Using playful but extreme textures and shapes, or sculpting and changing form around the body, are themes which permeate Miyake's uncompromisingly experimental work. The *Making Things* exhibition of 1998 showed the full scope of Miyake's canon: clothes burst forth from compressed gold foil capsules; sculptural, pleated forms collapsed like concertinas and then jumped in moving structures; selective chemical shrinking of deliberately oversized clothes produced crumpled surface textures; and an artist's performance of a gunpowder explosion left a random visual trail over pristine cream *Pleats Please* garments laid out on the floor.

The longevity of Issey Miyake's *Pleats Please* collection, since its start, has been due to its consummate wearability by a wide range of customers who defy categorisation and whose body shape may be far from the catwalk ideal. The ability of *Pleats Please* to flatter and enhance the wearer, through endless variations in shape on the body, its comfort and its easycare properties (handwash, no ironing) have sealed its success with professional women. The 100 per cent polyester fabric enables the increasingly complex pleat patterns to be heatset after the clothes are made up, creating a permanent effect. This is similar in concept to the 'drip dry' synthetics of the 1950s and 60s, but its aesthetic superiority has redeemed the status of polyester in common perception. The continuity of fabric and technique, and the timelessness of the concept of *Pleats Please*, positions it outside fashion in terms of radical

Kenzo
The Itineraries
Autumn/Winter 2004–2005
photo: Patrice Stable

changes in silhouette; however, the shapes and colours are constantly updated with new print treatments (such as harlequin designs or kitchen implements) and new pleat geometries (for example diagonal or two way pleats) and, more recently, with laser-cut lace effects, all of which are worked onto scaled up garments before the pleats are set.

Miyake himself is now concentrating on his ground-breaking *APOC* collections (the acronym derived from A Piece of Cloth) together with designer and fabric technologist Dai Fujiwara. After 15 collections, the radical concept of complete garments emanating from a single thread has been realised in a variety of ways. The first *APOC* collections were knitted using specially adapted Raschel knitting machinery, to create a wardrobe which is cut out from the tubular mesh knit, leaving little waste. Playful geometries have been developed, such as circular designs, and

the range now comprises both simple basics (termed *Baguette*, like the bread, which can be cut anywhere) and increasingly intricate patterning within the knitted cloth. Since 2000, when the first woven *APOC* double cloth garments without seams were devised (named *Pain de Mie*), continuous development work has resulted in highly complex double weaves. For example, the complete cutting pattern of all 27 pieces for a denim jacket is woven into the cloth itself, waiting to be released and sewn together; or the recent double layered coloured cotton cloth tile transforms into a tailored jacket. This process represents radical new ways of manufacturing clothes, where the fabrication, design and construction fuse into one, a new paradigm for clothing production.

After 30 years of inventive fabric developments with Issey Miyake, Makiko Minigawa launched her own clothing label Haat in 2000, under the Miyake Design Studio umbrella, incorporating both Indian and Japanese textile production. Her fabrics have

been shown in exhibitions in America and Japan, but they come into their own when used in the clothing, where the textiles take precedence over the cut. The clothing is based on materials and techniques from around the world, strongly influenced by traditional Indian and Chinese costume with its hand-stitched quilting and corded fastenings, but is coupled with modern knitwear, manufactured with advanced technology. For Spring/Summer 2006, inspiration was taken from the French-speaking Sahara, but the collection represents a fusion of influences and craft skills and is not intended to be identified with one country—a truly borderless fashion. Fabrics range from shrink-resist bubbly textures, hand-drawn prints, and swirling metallic embroideries, thick with surface texture or translucent, using fine cotton

or ultra fine nylon; with cut-out patterns, appliqué ribbons, hand-staining and *shibori* dyeing, the maker's hand is always in evidence. Nuno Corporation, from Japan, are internationally renowned for their highly innovative and experimental woven and constructed fabrics, which combine tradition with new technology—for example metallised glowing surfaces, crumpled textures, and unconventional treatments, including the use of rusty nails to stain the textile surface, or selectively shrinking fabrics using salt. Materials combine natural with synthetic fibres, as in jacquard woven with polyester and *washi* paper, while processes include spattering metals onto a polyester cloth, 'burn-out' or *devoré*, which eats away part of the surface, or embroidering onto soluble

substrates for lace-like effects. Nuno's work infuses the highest level of designer fashion, including early Issey Miyake, Comme des Garçons, Donna Karan, Calvin Klein in the USA, ATO, Cerrutti, and many more.

Nuno was started in 1984 by the visionary textile designer Junichi Arai, and is now under the direction of his collaborator Reiko Sudo with a small team of 12 designers. The textiles are highly acclaimed, having been celebrated in exhibitions in America and Europe, and promoted as part of 2005 Tokyo Fashion Week.[5] The unique qualities of Nuno textiles lie in an extraordinary and imaginative creativity in both concepts and materials, combined with a commercial (yet still exclusive) scale of production, utilising the long-standing history and artisanal skills of Japanese weavers. A set of books has been produced by Nuno which beautifully illustrate their philosophy and fabric themes, including *Suke Suke* (sheer, flimsy, translucent) and *Kiri Kiri* (glittering, brilliant, resplendent).

Experimental designer Yoshiki Hishinuma began by working for Miyake, but soon developed his own label, specialising in fabric treatments and technical innovations including crushed, printed and heat-pressed polyester, and layered laser-cut fabrics which are arresting in their complexity and splendour. Like Miyake, basic garments are made up oversized, then fabrics are machine stitched and gathered randomly, before being placed under a heavy transfer press with a pre-prepared image on special paper which is transferred to the cloth by heat, simultaneously setting the crumpled texture flat. One key feature of this process is that 'shadows'— hidden areas which do not receive the image— are created by creases, resulting in a unique

and random aesthetic. Recent collections by Marithe and Francois Girbaud, amongst others, have used this technique.

In Paris, new couturiers have emerged in recent years to challenge the decline of couture. In 1987, Christian Lacroix was the first designer to be admitted to the Chambre Syndicale for more than 25 years. Lacroix is known for his richly embellished fabrics, provocative silhouettes and his extravagant mix of patterns and textures. Jean Paul Gaultier, once the *enfant terrible* of French fashion, famous for his gender-bending, fetishistic 'underwear as outerwear', opened his own couture label in 1999 and joined the ranks of the establishment. His use of textiles has always been experimental and includes many distinctive allover prints; he was among the first to adopt digital printing technology on fine stretch jersey. He has also

top
Yoshiki Hishinuma
Autumn/Winter 1999
heat transfer print on polyester
photo: Sion Parkinson
courtesy Sandy Black

bottom
Jean-Paul Gaultier
Autumn/Winter 1998
photo: Anthea Simms

used hand-knitting to great effect, sending up the classic Aran knit with his wedding couple of 1998, dressed head to toe in cable knitting. At the other extreme, his couture collections have included stunningly beautiful hand-knitted and crocheted dresses, one of which was exhibited at the Victoria and Albert Museum in the *Radical Fashion* exhibition in 2002.

Over the last 20 years, the influence of the textile innovations discussed above have permeated to other levels of fashion. The scope of materials now utilised in mainstream textiles for fashion has become very broad and experimental; treatments such as felting and tie-dyeing, previously only feasible for individual items, are now available on a commercial scale on the high street. It has been possible to find the stretchy egg-box textured *shibori* tops, previously exclusive to high fashion designer outlets, being sold very cheaply on market stalls from Florence to London. The striking 'bleeding' stripe effects of Prada's skirts, cardigans and trousers for Spring/Summer 2004, created using a dip-dye technique, also quickly found their way to the mass market.

Similarly, the advances in technology of knitting have led to much more complex techniques, shapes and structures becoming widely available. Family dynasty Missoni have undoubtedly become the most well-known designer company for knits over the course of their 50 year history. Through their vibrant and colourful work, the industry began to see knitwear in a new, more fashionable light, and Milan gradually became the new fashion centre of Italy, encouraged by Italian post war reconstruction and promotions in American *Vogue*. Missoni is unique in its use of colour and pattern for innovative knit structures,

which are completely integrated with the garment design, and co-ordinated with prints for a signature 'total look'. They use both old and new knitting technologies, including old warp knitting machines to create their well-known zig-zag stripes, and modern weft knitting machinery for complex and large-scale jacquards. Being technologically harder to imitate, the Missoni look remains completely distinctive. The children of the family are now taking the collections for men and women into the younger fashion market.

A key development in commercial knitting was the arrival of Lycra and elastomeric yarns, which had a great impact on knitted jersey fabrics, enabling stretch to impart a more body-conscious silhouette, and improving fit for hosiery and sportswear. Azzedine Alaia, the

Missoni
Spring/Summer 2006
photo: Anthea Simms

Tunisian born designer based in Paris, was one of the first fashion designers to recognise the potential for knitwear to become sexy, utilising a high percentage of Lycra for his stretchy knitted fabrics in leopard skin patterns, velvet textures and supple viscose, then sculpting and moulding them to closely fit the bodies of his celebrity and model clients, simultaneously concealing and revealing the body's form. In the mid-1990s, knitwear traditions were again overturned when Julien Macdonald created his seminal cobweb lace dresses for Karl Lagerfeld and Chanel, followed by collections under his own name that led to his dressing glamorous celebrities in revealing outfits, often encrusted with antique beads, sequins and lace. Spectacularly, in 2001, Macdonald showed a dress embellished with a million pounds' worth of diamonds.

If Missoni established the new style in knitwear in the 1950s, this was paralleled by the extravagant colourful prints of Emilio Pucci's trend-setting modern fashion, using the newest synthetic fibres for optimistic, lightweight and carefree clothing that was rapidly taken up by the 'jet set'. Of note is the frequent use of allover large-scale swirling patterning fitted to the simple garment shape ('engineered' prints), which was at the time screen printed, but would no doubt be achieved by digital printing technology today. The lasting relevance and modernity of Pucci's signature multi-coloured, and often psychedelic, print aesthetic of the 1960s has influenced designers of today as diverse as Zandra Rhodes, who developed her own inimitable decorative 'handwriting' in the 1960s, Matthew Williamson, and Jonathan Saunders, whose debut collection in 2003 echoed the designs of the master, but with

the freshness of a new century's perspective. Matthew Williamson studied textile print at Central St Martins, launching his own style of colourful and bold patterning in 1997, and developed a personal and eclectic mix of graphic and representational imagery. His multi-hued work recalls that of Pucci and the great Italian textile houses. Often called 'hippy chic' style, he mixes prints in *ombré* rainbow stripe effects with colourful Indian embroideries, using images from Hindu divinities and exotic birds to feathers and cobwebs. The circle has now been squared, as Williamson was appointed in September 2005 to be the designer of the contemporary Emilio Pucci range, a line continued in the years following Pucci's death by his daughter.

The gamut of printed textile design is broad, as can be seen in the following pages. Many print techniques use special chemicals to

detail of *devoré* velvet fabric, c 1925
photo: Sion Parkinson
courtesy Sandy Black

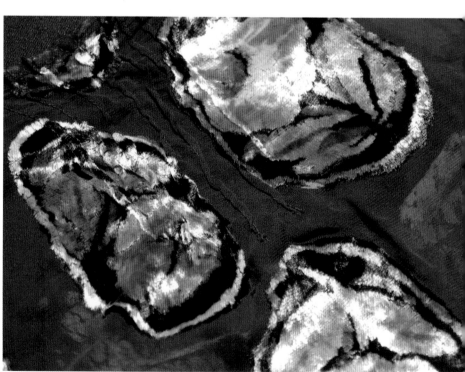

create not only pattern, but sculptural surfaces, such as the *devoré* process which burns fabric away, or printing with heat-reactive materials which form relief patterns. However, the mark of the artist's hand has been a crucial feature in printed fabrics—with painting and drawing having been the cornerstone of the designer's art until the advent of computers opened up the use of more photographic and graphic styles (of which the industry had been highly suspicious until recently), as well as creating new possibilities for production, as design work can be digitally transmitted. Semi-realistic floral designs still represent the bulk of the mainstream printed textile industry—a motif that is conspicuously absent in much contemporary designer fashion.

In this context, the work of Trinidad-born Althea McNish broke new ground in the 1950s and 60s with her vibrant, bold and painterly designs, abstracted from flowers or using irregular rhythms of pattern, which were commissioned by Liberty and Ascher in London, and by Balenciaga, Dior and Cardin in Paris. In contrast, Celia Birtwell's quirky hand-drawn textiles perfectly complemented the fluid crêpe viscose designs of partner Ossie Clark, whose fashions embodied the 'swinging London' of the mid-1960s. True to the cyclical nature of fashion, Birtwell's designs are once again popular, this time in her own name.

During the 1980s, print design was out of favour in fashion; however, the following decade saw the emergence of a new generation of designers creating bold prints. Ann-Louise Roswald, for example, produces large-scale graphic designs in two or three strong colours, printed on sweaters, skirts,

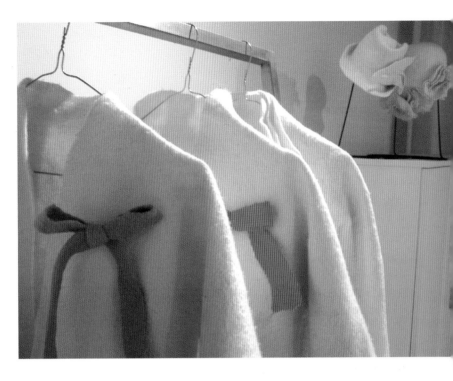

stockings and coats for a 'total look'. Eley Kishimoto, who started out contributing textile special effects to designers such as Hussein Chalayan—for whom they created the 'pixel flower' engineered print in his 1996 *Nothing/Interscope* collection—unusually went on to establish their own fashion label.

Somewhat surprisingly, much fashion and textiles education has a tendency to maintain a separation between the two areas, with clear lines of demarcation. London-based designer Shelley Fox's individualistic take on fashion is perhaps informed by her unusual training in both textiles and fashion design, which has led to her distinctive forms and often destructive textile treatments, such as blow torching sequinned fabric, or singeing textiles. When a designer constructs the fabric directly, as in knitting or felting, the development of fashion becomes a hybrid, combining fashion and textiles design, where form can be simultaneously created by the fabrication

Christine Birkle
Hut Up Berlin
Autumn/Winter 2006

itself. This is seen clearly in the work of Anne Maj Nafer and Christine Birkle, who work with shaping felted fibres and fabrics, and Delphine Wilson's work in hand-knitting. The changing cycles of fashion, and the rapid spread of ideas, stimulate desire over necessity, sustaining a constant flow of demand to be supplied. Once precious silks were brought to the West from the East as trophies of arduous journeys to far-off lands. Now the ease of international travel and satellite communication has realised the 'global village', a cultural melting pot from which new and hybrid ideas are emerging, permeating all levels of creative activity from

food and music to new fashions. Traditional textiles and costume from around the world, particularly from Asia and South America, are regularly referenced in Western fashion design. For example, in the 1980s, Kenzo recreated versions of the gathered skirts worn by the South American *cholitas*, and Romeo Gigli echoed the opulence of Russia's czars; John Galliano's thematic collections in the 1990s presented at times the American Indian or the modern Geisha. With increased globalisation and cultural crossover, in more contemporary fashion, traditional costume is combined in new ways, sometimes with political or religious overtones. Hussein Chalayan references trans-cultural identities and nation states in his work, offering an eloquent commentary on the current political and religious climate by showing powerful sequences of transformation. For example, in *Between*, Spring/Summer 1998, he showed a series of models in *chadors*, decreasing in length, from totally covered to totally naked, save for a facial mask. In *Ambimorphous*, Autumn/Winter 2000, Chalayan deconstructed traditional Turkish costume to gradually merge it with Western black dress.

A wave of experimentation and spectacle in the mid-1990s focused on more and more elaborate and provocative fashion shows, especially from London designers Alexander McQueen, Hussein Chalayan, Julien MacDonald, and John Galliano, but also by Jean Paul Gaultier, Walter von Bierendonck, and Thierry Mugler in Europe, and Alexander Herchcovitch in Brazil, among others.[7] These occasions provided catwalk as theatre, concocting elaborate scenarios, one-off clothes and extraordinary fabrications, many of which have become seminal moments in fashion history. Chalayan, always a

John Galliano
Autumn/Winter 2005
photo: Anthea Simms

conceptually driven designer, has concentrated on silhouette, materials and precisely engineered transformations for breathtaking theatrical impact; for example the furniture which became clothing and portable environments in 2000's *After Words* collection, shown in a theatre space. In 1994 he transposed the tough but paper-like packaging material Tyvek into 'airmail clothing'—dresses and suits that could be unfolded from an airmail envelope arriving through the post.

Meanwhile, Alexander McQueen's dark, thematically styled visions challenged the audience and the press, whilst also showing beautiful, disturbing and romantic clothes in exquisite and experimental fabrics (many especially developed for the shows). For example, in McQueen's 1993 *Nihilism* collection, clothes were coated to look blood-stained, or printed with tyre tread patterns; in *Dante*, 1996, delicate laces were fashioned to the precise but extreme shapes of the garments.

McQueen and the other spectacular showmen designers commission exceptional fabrics for these shows, providing a platform for many textile and accessory designers. One result

Hussein Chalayan
Between
Spring/Summer 1998
photo: Chris Moore

of this media-focused activity was the international attention given to visionary designers, particularly by the established fashion houses of Paris, as the y recognised that couture could benefit from a new injection of radical approaches and youthful ideas. This resulted in landmark appointments of British designers at Givenchy (John Galliano, Alexander McQueen, Julien MacDonald), Dior (John Galliano), Cacharel (Clements Ribeiro) and Chloe (Stella McCartney, Phoebe Philo). These young designers thereby gained a financial security not otherwise available to them, whilst retaining for themselves a significant amount of creative freedom. The survival of the design houses, in turn, was assured, whilst brand building and acquisition culture continues. The quest for ever more elaborate textiles for one-off showpieces to fulfil the catwalk scenario leads to innovatory pieces which, with the all-important financial backing, can be converted into an approximate commercial version for the industry.[8]

New fashion centres are emerging throughout the world as developing economies in China, Malaysia and India start to compete in the retail brands arena, and begin to move from being mainly textile producers serving the Western fashion system to establishing their own vertically branded businesses. In tandem with this shift of emphasis, a new generation of fashion designers arising from these regions is beginning to impact on the international fashion world—such as Manish Arora from India and Márcia Ganem from Brazil. Such designers express, through their collections, a desire to utilise the best of traditional skills and influences and absorb these into multi-culturally inspired ranges of clothes, thus bringing these specific heritages into the fashion arena.

Some initiatives have begun to recognise the importance of self-sufficiency and the preservation of traditional livelihoods. International model turned designer Bibi Russell returned to her native Bangladesh in 1994 with the intention of creating new Western markets for traditional hand-woven and embroidered cloths. Russell's Fashion for Development mission, supported by UNESCO, aimed to combat poverty, and to gain international recognition for Bangladeshi craftsmanship and traditional fabrics such as *khadi* (handspun and hand-woven cotton), *dobi* (striped cotton), and the complex *jamdari* (silk cotton gauze), which takes one month to produce six metres. This work is not about fashion in the European sense—but about the livelihood of a people and the

Bibi Russell
Spring/Summer 1997

promotion and celebration of a country which is often only associated with the misery of floods and other disasters. This pioneering work is now continued by ethically-inspired labels like Juste, a London company which uses textiles sourced from Bangladeshi producers mixed with classic fabrics for a young fashion look.

Other companies, based in India, such as Anohki and Raag, have carefully worked to preserve the value of craftsmanship, supplying the American and UK markets with hand-woven textile production and hand-dyed or embellished fabrics and clothing. The intricate tie-dyeing work and *kabira* hand-stitched surface textures and quilting of the Raag workshops can also be seen in the collections of Issey Miyake and Haat. Long established and highly successful Indian designer Ritu Kumar aptly describes the rich textile legacy of India residing "not in museums but in the hands of its surviving 16 million artisans".[9] She has revived and updated traditional blockprints and *khadi* and *bandhani* fabrics for a modern clientele, whilst maintaining an authentic identity. This reservoir of skills is also magnificently utilised by Abu Jani and Sandeep Khosla, couture costumiers for film stars and prestigious customers, in their sumptuous outfits, embroidered and encrusted with beads.

The designers featured in *Fashioning Fabrics* are diverse in both the fabrication techniques they employ and their years of experience in the fashion business. All have been selected as textile-led designers, who initiate the development of original surface textures, structural form or pattern effects, sometimes personally crafting their fabrics (particularly in the case of younger designers), and

combine these with silhouettes and cutting to create the overall impact of a collection.

This book attempts to acknowledge the great wealth of talent currently working in this way within the international fashion arena, and if it serves to showcase the hidden arts of textiles used in fashion then its mission is accomplished.

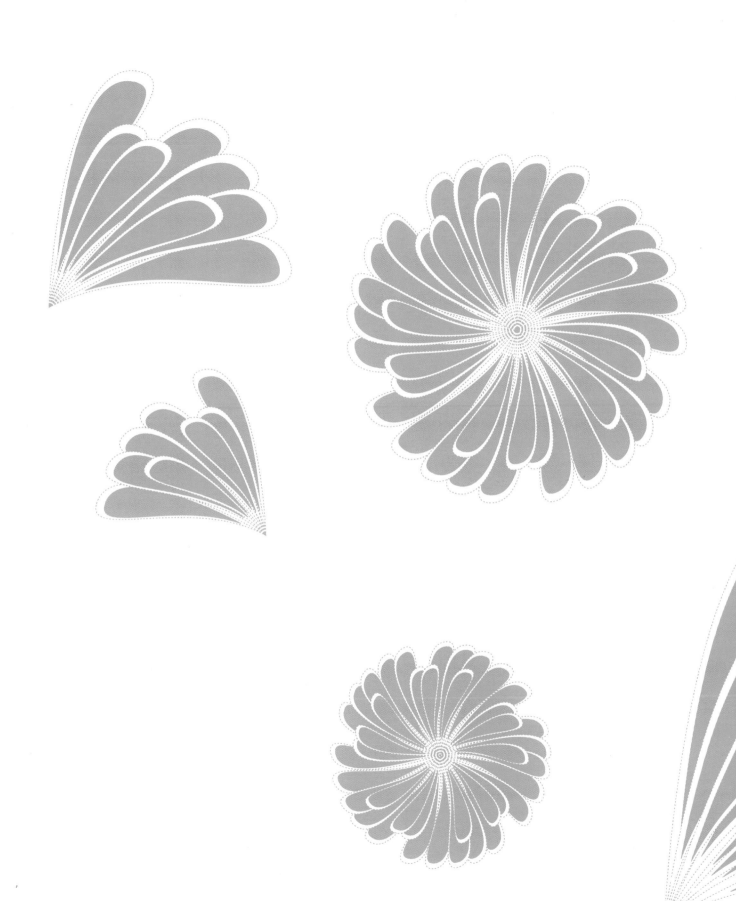

embellished

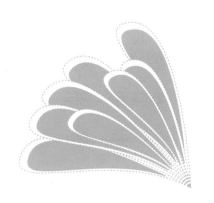

stephanie**aman**

Generally interpreted as 'embroidery with an edge', Stephanie Aman's designs re-contextualise the antique art of embroidery alongside distinctively modern cuts in natural fabrics—a signature fusion that has seen the young designer attain a foothold within the ranks of the competitive British fashion industry. Recognised for her use of contrasting fabrics, detailing and intricate handcrafting methods, Aman's challenging garments play on the unique tension occurring through the synthesis of divergent, yet complementary, materials. Combining leather and silk chiffon in hues ranging from warm nudes and yellows to muted charcoals, Aman is interested in "the contrast in qualities of the hard and soft fabrics".[1]

Aman's predilection for these contrasts is most evident in the meticulously beaded gun-metal grey neck and shoulder adornments throughout her range—an inventive design feature that creates an almost Grecian armour effect. This effect is heightened when coupled with the delicate layers of embroidered silk chiffon in regal hues of crimson and gold that billow cape-like from the shoulder blades. Elsewhere heavily embroidered grey silk hoods and black leather smocks fixed with silver buckled straps allude once again to ancient battle costume, offset by delicate layers of silk.

Each garment is created through an organic process; says Aman, "I allow the fabric and embroidery to develop and evolve naturally. The embroidered style often creates the look of antique lace, it has a feeling of age and history but used in a modern context." Not designed to be mass-produced, Aman's garments are inherently individual—no two pieces can be the same due to the nature of their production.

Aman's work is all the product of free machine embroidery, "I draw with the needle", Aman explains. "I treat the threads as my colour palette and the fabric as my canvas." At this early stage of her career, London-based Aman has managed to carve out a name for herself as a designer who takes traditional methods and materials and steers them in new directions. "My strengths lie in working from research and translating this to embroidery and garment design. Couture and fine craftsmanship can be the essence of any collection, but the details of colour, fabric research and silhouette are just as important in their own right."

Indigo Clarke

Dum Spiro Spero
(While I Breath I Hope)
Royal College of Art
graduation collection, 2005
photo: Caroline Marks

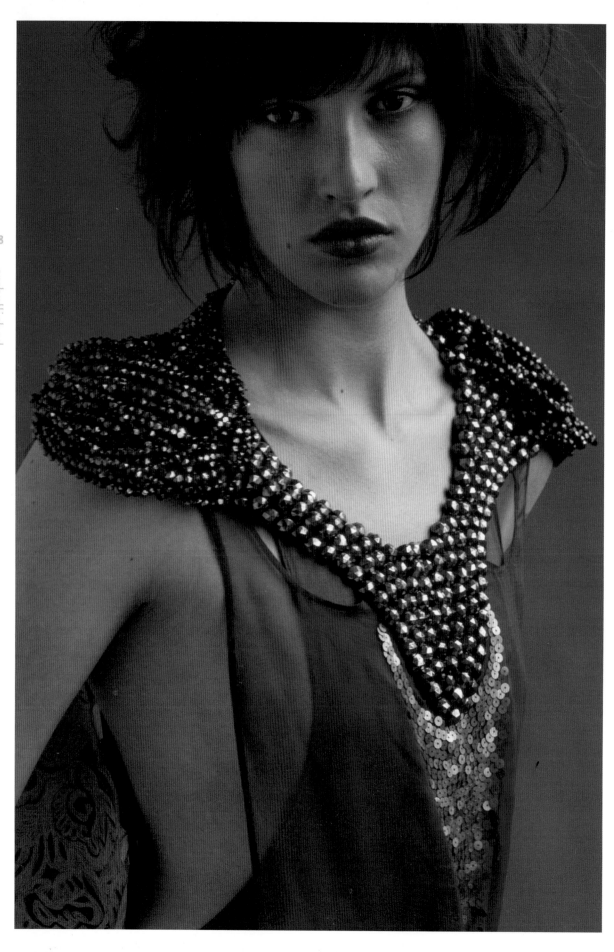

opposite
Dum Spiro Spero
(While I Breath I Hope)
Royal College of Art
graduation collection, 2005
photo: Caroline Marks

manish**arora**

Manish Arora emerged on the London runway in late 2005 with a masterful fusion of Western modishness and native Indian symbolism, all manifest in a lavish array of textiles. Arora's early collections, dating back to a graduate collection for New Delhi's National Institute of Fashion Technology in 1994, featured an incongruous tug-of-war between Indian and Western iconography. Tee-shirts screen printed with Lakshmi, the Hindu divinity of wealth, beauty and good fortune were paired with skirts alternately printed with "Jack and Jill Had Sex" and appliquéd with swinger-style dice. Arora's ability to intertwine these conflicting aesthetic histories has since matured, and his employ of more complex, refined, and historically informed textiles correlates with this evolution.

Manish Arora's Spring/Summer 2006 collection incorporated a number of traditional Indian embellishment techniques, all executed on the midriff tops, fitted boleros, and full Romantic sway skirts currently popular in Western fashion. The opulence of his label derives from a breadth of vivacious fabrics that, through a determined revival of Indian textile crafts, beget surprisingly contemporary garments. Several of the season's designs championed the *zardozi* embroidery technique that is used on silk sari fabric. *Zardozi* is a Persian tradition of fine giltwork stitching and can be literally translated as "gold" (*zar*)

"embroidery" (*dozi*). More lavish ensembles included a full cape of cut strips made to look like the regal plumes of the peacock, which has significant religious symbolisms in its native habitats of India and Sri Lanka, as well as a jacket of polychrome textured silk edged with *chikan* style embroidery, a technique qualified by the use of untwisted white cotton, silk, or more recently, rayon thread stitching.

Arora also created a series of vibrant silk skirts, impressively cut in a *Rajasthani* circular style and embroidered with stylised lotus flower, foliate and cloud motifs. Their matching bolero jackets boasted flame-like projections at each shoulder to configure a silhouette markedly inspired by Thai festival costume, and were often embroidered in the *shisha* or mirror work popular in India in the seventeenth century. Whilst the tiny mirrors are anchored to each jacket with a chain stitch, the more figurative embroideries utilised satin stitch, which is thought to have infiltrated the Indian textile market via Europe in the nineteenth century and overtaken native feather, herringbone, and chain stitch techniques. Through dazzling interplays between indigenous Indian fabric embellishments and charismatic couture-quality patterning, Manish Arora has carved his own niche in global fashion.

Elyssa Da Cruz

opposite
detail
Spring/Summer 2006
silk woven with metallic
thread, gold PVC cut-out
appliqué, metal and glass
beaded arabesques applied
by hand

opposite
Spring/Summer 2006
multi-colour woven silk
and metallic thread
jacquards, hand-sequinned
plain weaves, skirts with
appliqué and figurative
satin stitch embroideries,
graphic prints for body
stocking and leggings

detail
Spring/Summer 2006
silk woven with metallic
thread, gold PVC cut-out
appliqué, metal and glass
beaded arabesques applied
by hand

ashish

Featuring equal portions of eccentricity and glamour, Ashish Gupta's inimitable designs intertwine the seemingly paradoxical to grand effect. Combining contrasting influences such as Eastern and Western cultures, leisurewear and black-tie, traditional and contemporary design with high-quality textiles and Indian hand-crafting techniques, Gupta has emerged as a leading figure in the new generation of British fashion.

Gupta's sense of irony, his fearless merging of audacious prints with intense colour schemes, intricate detailing, as well as consistently high standards of craftsmanship and materials, have received international acclaim since his 2001 debut collection. In response to disposable 'high-street' fashion, Gupta exclusively uses fabric he has designed himself—not only rendering each item unique, but taking historic and often overlooked crafts into high fashion. "I feel it is so important to keep craft traditions and techniques alive, as in India they are passed on from generation to generation", Gupta explains. "This is especially important in an industry which is more and more reliant on mass-manufactured clothing. Creating quality, hand-crafted garments is a statement against mass-production and big corporations."[1]

Gupta adapts techniques, such as weaving and embroidery, to create complex patterns and unique fabric designs. What may begin as simple cotton cloth will be completely transformed once another fabric, such as lace, is appliquéd onto it, re-embroidered with wool, and covered with sequins. Any one of Gupta's intricately produced garments could have up to three people working on it, as the embroiderers each specialise in different forms of embellishment.

Countless references to archetypal designs litter Gupta's collections, distorted and re-interpreted through the way in which the textiles have been treated. His shimmering, sequined sportswear sets for men and women, and 1950s inspired sunflower appliqué prom dresses and crop jackets, showcase the Indian hand-work whilst paying homage to classic formal and casual cuts—with a tongue in cheek injection of the burlesque. The Spring/Summer 2006 collection included sequin-encrusted Rugby shirts luxuriously recreated out of silk georgette, sequined 80s inspired multi-coloured shirts, casual dresses and men's summer suits in traditional Afghani *ikat* inspired fabric, hand-woven out of pure cotton yarn. "I want my designs to be really beautiful and timeless, and to go beyond the whole 'trend' thing", says Gupta of his original garments. "It is more important now than ever for designers to make something unique, something precious that cannot be copied and mass-produced."

IC

top
detail
Autumn/Winter 2005
wool and sequin embroidery
on cotton

bottom
detail
Autumn/Winter 2006
wool embroidery on
wool crêpe

left
Spring/Summer 2006 hand-
woven cotton *ikat* jacket
and shorts with sequin
hand-embroidery, and
hand-woven cotton
ikat shirt
photo: Chris Moore

right
Autumn/Winter 2005
wool and sequin hand-
embroidered coat on cotton
fabric base
photo: Ian Gillett

opposite
Autumn/Winter 2006
wool crêpe dress with wool
hand-embroidered flowers
photo: Chris Moore

ioannis**dimitrousis**

embellished

The words brightly coloured, multi-textured, patched, fringed, and deconstructed best summarise the work of Greek born designer Ioannis Dimitrousis. Following his first professional debut as part of London Fashion Week's On/Off schedule in February 2006, the designer now runs his own label from his studio in London.

Unexpected fabric mixes and decorative effects, especially in menswear, have already become a signature of the young designer. Layers of vintage fabrics and knitwear are given a new lease of life through his interpretation of recycling, mixing the old with the new for his sportswear inspired look: "I wanted to create a mass of fabric around the wearer, not to feel trapped, but secure and warm." Dimitrousis creates bespoke garments which challenge proportion by combining oversized accessories with neatly tailored jackets and casual separates. Chunky crochet openwork scarves, hand-made by the designer's mother, combine a multitude of colours, hundreds of pearls and vintage fabrics to create his tactile and colourful pieces.

One piece features an intense area of multi-coloured beadwork—again by his skilful mother—overlapping the side seams of a patchwork fabric top, creating a contrast of opulence with the mundane. His fabric treatments and juxtapositioning, including

his recycled graduate collection, can only be described as eclectic. The 25-year-old's sartorial expertise is a product of his illustrious education and work experience which included pattern-cutting and assistance to some of London's top designers, including Louis de Gama, Roland Mouret and Jonathan Saunders.

While the label has only been launched since October 2005, Dimitrousis has already established quite a following, especially at home in Greece, where celebrity fans include a famous Budho dancer, Gabrielle Daris.

Ellie Rivers

Autumn/Winter 2005
London College of Fashion
graduation collection
tailored jacket made from
natural denim fabric,
hand-embroidered with
pearl beading

johngalliano

A fierce subscriber to the ocular amalgams of postmodernism in fashion, John Galliano has established himself as a master of the theatrical, both under his own label and for the formidable House of Dior, for which he was named Creative Director in 1997. The conductor of an orchestra of collage, each Galliano creation forms a visual synthesis of textile and constructive elements that derive from a range of eras and cultures as variant as the Belle Epoque's *demi-mondaine* and the Maasai warrior tribes of Africa.

His work reflects equal attention to the subversive, whether revolutionary or rebellious, and to tireless self-promotion. In his Autumn/Winter 2005 exposition, the two fused in bias-cut trenchcoats and dresses of cotton jersey and Teflon with a "Galliano Gazette" allover newsprint; each of these newspaper concoctions was tagged with fluorescent and black haphazard graffiti renderings. The French Revolution has creatively inspired Galliano more than once, titling his 1983 graduation collection from Central St Martins College of Art and Design *Les Incroyables*, and emerging recently in his work for the Spring/Summer 2006 Christian Dior Couture presentation. The designer acknowledged the protests and revolutionary tidings that riddled Paris throughout 2005 and early 2006, and translated this contemporary conflict into red discharge-dyed ponyskin

tailored jackets and white organza and chiffon chemise gowns, some of which were embellished with 'bloody' polyamide splatters and silkscreen prints of Marie Antoinette.

As far-reaching and elaborate as some of his thematics tend to be, Galliano is brand-savvy and business-minded, and invests in both of his labels. Galliano has visited many of the crinoline factories and textile companies that Dior himself patronised in the 1950s, and consistently attempts to reference mid-twentieth century image-makers and Christian Dior confidantes like Cecil Beaton, Irving Penn, and Rene Gruau, as well as muses Zizi Jeanmaire and Wallis Simpson, in dramatic translations of classics like the Bar suit of 1947's New Look, or the asymmetric 1948 Envol line. For his first collection for the House of Dior, John Galliano affected the wasp-waist bodices and rounded hiplines of Dior's heyday, but with Dinka corsets and Maasai jewellery of multi-coloured beadwork, and then situated them atop heavy taffeta bustle-back skirts redolent of nineteenth century portraiture.

John Galliano's tendency towards overwrought silhouettes predicates extravagant fabrications, made opulent by beadwork and silk floss embroidery, and by generous applications of luxe raw materials. His chic Spring/Summer 2006 ready-to-wear line, cast an alternative set of giants, dwarfs,

Spring/Summer 2006
silk crêpe with silk floss
embroidery and polka-dot
and floral printed silk faille
photo: Anthea Simms

twins and magicians, and was expertly finished in floral and jewel encrusted embroideries of demure petal-shaped tiers, forming elegant flapper dresses and slim-cut ballgowns. The designer has used a wide range of furs for trimmings, accessories, and even full coats over the past several years to great effect, most notably in his Autumn/Winter 2005 collection, which offered Op Art-style jackets inset with swansdown and fox fur, 1930s-style pink maribou negligees with gilded four-leaf clover appliqués and Paul Poiret-esque hobble gowns with dyed yellow coyote fur cowlneck collars. Galliano is equally well-versed in his applications of fine leathers and feathers. In his Autumn/Winter 2002 presentation for Christian Dior Couture, a flamenco-inspired swing coat of navy parachute revealed an underbelly of plumage with the model's twirl, the skirt lined heavily in ostrich feathers. Beneath, a sinuous sheath of black silk tulle, appliquéd with laser-cut squares of crocodile leather to re-affect the reptile's skin, showcased Galliano's skill in the manipulation of animal spoils.

Long a proponent of deconstruction as a means to authenticity and quality in high fashion, John Galliano worked as an assistant to the reconstructionist tailor Tommy Nutter in London in the early 1980s, and has since incorporated dissemblage as a crucial part of his design vocabulary. One ensemble, produced for the Autumn/Winter 1996 runway, was constructed from a beige suede, violently laser-slashed and clipped into a fringe along an asymmetrical hem. Finished with frilly scalloping and laser-cut shapes reminiscent of traditional western cotton undergarments, the dress, both in its transparency and its liquid drape, embodies the infra-apparel mode of the late twentieth century. The garment was paired with a white wool cocoon coat printed with *ikat* patterning—a pre-weave dyeing technique cultivated indigenously in Java and South and Central America—which lent the ensemble an exotic tone and a dense aesthetic framework.

Galliano's eclectic fusion of embellishment techniques, and visual referencing both staunchly historical and playfully socio-cultural, provide him a distinct personality in high fashion. Of Spanish descent, the London-raised, Paris-dwelling *createur* has said: "I think all that—the souks, the markets, woven fabrics, the carpets, the smells, the herbs, the Mediterranean colour, is where my love of textiles comes from."

EDC

opposite
Spring/Summer 2006
digitally printed silk
with silver *paillette* and
rhinestone embroidery; silk
tulle with gold bugle bead
and *paillette* embroidery
photo: Anthea Simms

opposite
Autumn/Winter 2005
silk chiffon overlay and
gold-leaf embossed four-
leaf clovers with
maribou trim
photo: Anthea Simms

Autumn/Winter 1996
ikat-printed felted wool
with silk lining with tartan
wool twill
photo: Anthea Simms

yoshiki**hishinuma**

As a former apprentice to Issey Miyake, Yoshiki Hishinuma's seamless conflation of traditional Japanese textile dyeing, sculpting techniques and cutting edge developments in fibre and processing is hardly surprising. With the introduction of his own fashion label in 1992, Hishinuma has exhibited a fondness for various synthetic fibres, most notably polyester chiffons and organzas. His use of polyester, as well as raw cotton, synthetic velvet, and woven metal textiles serves to provide his architectural mouldings with intricate textures and surreal projections. The designer's fabrics are almost exclusively developed and manufactured in Japan and his garments are produced in an Aoyama atelier. In *Structure and Surface: Contemporary Japanese Textiles*, author Matilda McQuaid explains the popularity of polyester amongst the avant-garde Japanese textile set: "This prosaic fabric has been enlivened by texturing its surface, an approach often used to conceal defects in lesser-grade plastics or glass. Heating, steaming, puncturing, dissolving with acid, polishing, clipping, shaving—abusive treatments associated with durable materials like stone, ceramics, or glass—transform polyester into cloth that challenges our notion of what textiles can be."[1]

A prolific designer, Hishinuma has experimented with every tailoring technique from postmodern reconstruction to apocalyptic deconstruction, but it is the textiles that dominate his aesthetic and will surely secure his legacy as 'an artist of the cloth'. His collections have consistently utilised *shibori* binding, a method of heat-moulding, that was pioneered in large part by artist Yuh Okano for the Daito Pleats Company in Gumma, and is used to bind polyester around metal discs or domes. Hishinuma has alternately emphasised this egg carton-esque structure with bulbous knit jackets and tunics in Spring/Summer 2002, or ironed flat to affect the stiffened ripples or quilt compositions of Autumn/Winter 2002 and 2004.[2] Hishinuma's 1997 collection even featured hand-made Chinese textiles that were created by executing this process with metal baking trays to produce a fish-scale texture enhanced by the fluid range of electric blues inherent in the use of *shibori* binding.[3]

Shibori, an age-old Japanese resist-dye technique, has seen many manifestations in the Hishinuma portfolio, from the alien protrusion to more traditional clusters of blistered fabric that appeared regularly in the designer's contemporary kimono designs for the 1992 movie *Goh-hime*, directed by Hiroshi Teshigahara. And this technique was used further in Hishinuma's Spring/Summer 1999 collection to create almost every pattern imaginable with a vast array of complimentary electric hues.

opposite
Autumn/Winter 2001
laminated urethane forms

Yoshiki Hishinuma has worked extensively with the Japanese textile firm Nuno, headed by artist Reiko Sudo. His Spring/Summer 1997 collection featured polyester knits printed with multicoloured PVC (polyvinylchloride) bands. When the fabric was heat-treated, the polyester would shrink, forming wrinkles in the criss-crossing lines of PVC. This Nuno process lent itself to the tribal theme that governed Hishinuma's presentation by dressing face-painted models in figure-hugging sheaths, evoking the aesthetics of scarification.

This designer is well-versed in surface deconstruction as well, notably in the synthetic velvet *devoré* fabrics manifest in designs ranging from simple sheath gowns to impeccably tailored suits. Hishinuma's most dynamic deconstructive posturings feature a laser-cutting technique developed for his *Sand Planet* collection of Autumn/Winter 2002. One critic hailed the inspired presentation, writing that: "the designer drew on David Lynch's epic, *Dune*, to create a... collection centred on a strong and proud woman battling the fierce elements of an inhospitable desert landscape."[4] The garments, which featured bias drapes of PVC, slashed into fringes and looped strips, were as intrinsically traditional, with their kimono stylings and ancient Japanese court headdresses, as they were quintessentially contemporary. Hishinuma's

ability to render an impressive collage of the traditional and the avant-garde has infused his designs with an idiosyncratic quality, both chic and craft-like, that is unarguably reliant on expertly engineered textiles.

EDC

yoshikihishinuma

Autumn/Winter 2000
transfer printed
and urethane
laminated polyester

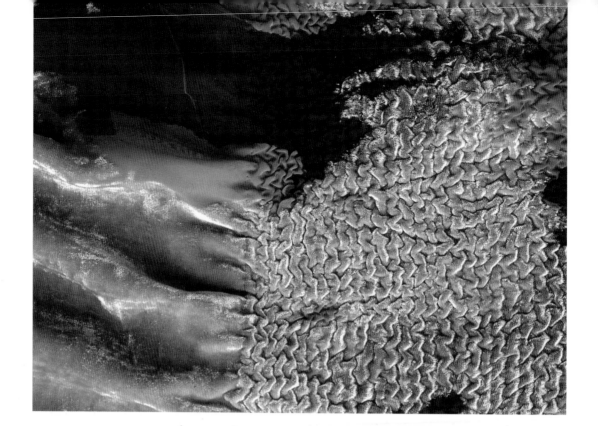

top
detail
Spring/Summer 1995
burn out and transfer print
on rayon/polyester velvet

bottom
detail
Autumn/Winter 2005
laser-cut rayon/polyester
velvet and synthetic leather

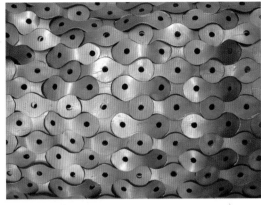

left
Antenna Jacket
Autumn/Winter 2003
hand-printed, shrunk wool

top right
detail
Autumn/Winter 2005
metal and transfer printed
synthetic leather

middle right
detail
Autumn/Winter 2000
shrunk and transfer printed
polyester and laser-cut leaf

bottom right
detail
Autumn/Winter 2005
needle punched wool

abu**jani**&
sandeep**khosla**

The names Abu Jani and Sandeep Khosla are synonymous with classic Indian style and elegantly draped silhouettes. Re-working the ancient arts of embroidery and beadwork into modern haute couture, Jani and Khosla have rejuvenated the Indian fashion scene by showcasing Indian textile embellishment techniques on a world stage. They are best known for reinventing heritage crafts like *chikan* and *zardozi* to create white on white embroidery, using untwisted threads of cotton or silk, and heavy needlework with silver and gold threads. The designers' creative use of sequins, pearls, and precious stones on rich textiles such as silk, velvet, organza, chiffon and *crêpe-de-chine*, have made their garments the choice of numerous celebrities from Bollywood and Hollywood alike.

Having started out as costume designers for the Bollywood film industry, Abu Jani and Sandeep Khosla incorporate their love of glamour and flamboyance into their garments by featuring imaginative embellishments, including Swarovski crystals and mirrors. Bollywood actress Tina Munim's wedding gown was an Abu Sandeep original comprising ten metres of maroon silk georgette embossed with patterns inspired by sumptuous Persian carpets. Abu Sandeep's flair for injecting eveningwear with the splendour reminiscent of the royal Maharajas has even landed their designs on the Oscars' red carpet, with a

delicate *dhaka* rose embroidered ankle length ivory tunic worn by Dame Judy Dench.

Using traditional crushed silk, crushed *bandhani* or tie-dye, and *dogri* sleeves, Abu Sandeep give a modern twist to classic Indian textiles. *Chikankari* embroidery, with its six basic stitches, is used to create imprinted shapes, and whilst the technique is normally reserved for summer attire, Abu Sandeep intersperse it throughout their formal eveningwear. *Zardozi*, the crafting of heavy embroidery using gold and silver threads on velvet and silk, is a specialised art that creates the richest detailing and is featured on a number of their luxurious bridal gowns and a popular line of wool scarves. After a decade of lavishness and vibrant use of colour and adornment in their design ethos, including lines inspired by traditional paintings, a more subdued collection emerged based on thread and bead embroidery. This called for the softer floral motifs of *dhaka*, and subdued colours like ivory and blue. The introduction of beads and sequins to the delicate thread embroidery straddles the line between their trademark heavy embellishments and the simplicity of *dhaka*.

Jennifer Trak

opposite
bridal dress created for Anju Chulani, 1993
ten metres of Benares silk organza embroidered with Swarovski crystals

Tina Munim's wedding dress
10 metres of pure silk
georgette velvet with
Persian carpet pattern in
pure brocade

opposite
a *zardozi* net sari

abujani&sandeepkhosla

kenzo

Details that beg, steal and borrow from across the globe are at the centre of Kenzo's design, creating a fusion of folklore styles that vibrate with a blend of fantasy and reality. Just as in the visual arts and literature, appropriation of styles and techniques from different locations, times and genres has become an increasingly important movement, so too has the craft of piracy been celebrated in fashion, and perhaps Kenzo has been the most influential pirate of them all, crisscrossing the globe and bringing clashing cultures together in single garments.

A Japanese man arriving in Paris in the 1970s, Takada Kenzo, founder of the House of Kenzo, began to design clothes at a time when fashion was still finding its limits. Spurred on by the youthful revolution in consumerism during the 1960s that was keen to go everywhere and try everything, Kenzo was liberated from the rigid European conventions of taste and style. His garments freed lines from the body, widening and loosening waists and arms, and fusing traditional suits with kimono sleeves brought from his native Japan. Standing at fashion's threshold between East and West, Kenzo appropriated and mis-appropriated styles from both traditions, with wit and style, in order to create an aesthetic that is globally appealing.

Kenzo has always been multicultural in the most appealing way, striking a fine balance between heterogeneity and unity in design. Takada Kenzo's embrace of influences from a wide variety of cultures, powered by their exuberant mastery of colour, pattern and texture, made the label a forum for diversity, with people from all ages and ethnic backgrounds appearing on the Kenzo catwalk —drawing together and celebrating difference just as he did in his designs. Strong patterns were often put together with other clashing patterns from different styles and periods, like tweeds with Japanese embroidery or jungle prints with bright tartans. What Kenzo has always managed to achieve in this fashion mishmash is a kind of poetic tapestry, a weaving together which is less the piracy of appropriation and more like the making of a patchwork memory quilt.

When Takada Kenzo retired in 1999, he left the label in the care of Giles Rosier, who was to remain fashion director until 2003, when he left to focus on his own label. When the Sardinian born Antonio Marras took over, he began to appropriate the Kenzo tapestry of styles in its most literal sense, piling layer upon layer of references, patterns and embellishments. For Marras' Autumn/Winter 2005 collection for Kenzo, details and styles from Inuit clothing danced happily with Russian Cossack skirts, gaudily coloured Scottish tartans, Japanese kimono sleeves and whimsical English florals, laced with glistening

opposite
An English Garden in India
Autumn/Winter 2005
wool and cotton tartan
jacket with wool skirt
embroidered with gold,
lurex and pearl threads
and sequins
photo: Patrice Stable

overleaf
detail from opposite image

golden Indian embroidery. Marras took his inspiration from travelling cultures for this collection, groups that are constantly on the move, carrying all their most important possessions on their person, even within the fabric of their clothes.

References are constantly in motion here, as they were in Takada Kenzo's work, interweaving space, time and place. As Marras explains of his nomadic muse: "On reaching your destination, it is time to start afresh, to be born and reborn, and to nostalgically delve into memories. Then once more you pack your bags, taking photographs, souvenirs, postcards and clothes with you. Clothes from different cultures and lands."

The clothes that appear in this 2005 collection for Kenzo, heavy with detail and resonance for the global nomad, are ultimately transformative, creating a new tradition of aesthetic from old references: a design folklore that will itself be plundered by fashion pirates in the future to come.

Laura McLean-Ferris

opposite left
An English Garden in India
Autumn/Winter 2005
embroidered wool cape with
taffeta dress
photo: Patrice Stable

opposite right
An English Garden in India
Autumn/Winter 2005
embroidered wool coat with
printed silk crêpe dress
photo: Patrice Stable

left
The Itineraries
Autumn/Winter 2004
multi-coloured boiled wool,
cotton, viscose and lurex
photo: Patrice Stable

right
The Itineraries
Autumn/Winter 2004
floral jacquard, boiled
lambswool and lurex with
crocheted flower bonnet
photo: Patrice Stable

christianlacroix

With 2006 marking his 25th year as a couturier, Christian Lacroix continues to rather remarkably intertwine his name with those of the most revered and luxurious textile houses in Europe. His acuity as a decorator is only surpassed by his skill as a historicist. Coupled, these qualities promote his brand as the perfect showcase for a range of textiles that includes sophisticated embroideries, elaborate crochets and lacework. Lacroix received the first of two De d'Or (Golden Thimble) awards from the Fédération Française de la Couture for his sculptural 'pouf' dress for Jean Patou, which utilised a doubled heavyweight silk taffeta, before opening his own salon in 1987.

Under his own label, Christian Lacroix has had the freedom to explore the influences of his Arlesienne roots, which aesthetically translate to textiles covered in patchwork, quilting, and a gypsy-esque fusion of various constructive and embellishment techniques, from stem-stitch silk floss embroidery to couched patterns of hammered gold and silver metallic bands and multi-coloured jewels. In fact, Lacroix's upbringing in the Arles region of southern France provided him with an eclecticism that has driven his fashion and textile choices for the entirety of his career to date. The designer explains, "Arles meant a world of artifice and baroque, of gentle and civilised folly and solemn futility."[1]

This "artifice and baroque" is conveyed through a focus on Spanish lacework, evocative of the textures and palettes of Francisco de Goya and Francisco de Zurburan, as well as the confectionary embroideries and floral printed silks of the eighteenth century. The former is informed by his obsession with the Nord Pinus, the bullfighter's hotel near Arles, and the latter fuelled by a youthful patronage of theatre and operettas, and by art historical studies at La Sorbonne in Paris.

Lacroix alternates between two Latino personas; the matador, tailored in rich black silk velvets with hammered and coiled gold threads, and the immaculate Spanish virgin, in white cotton organdie bejewelled with ruby and emerald embroidery. Both were configured through the textile artistry of the formidable embroidery house Lesage, which was founded in 1868. Lacroix's *quinceañera* and flamenco inspired headdresses, capelets, and gowns employ textiles from both individual designers and established textile houses. Ulrika Liljedahl, for instance, created a black machine lace and crochet dress, re-embroidered by hand for Lacroix's Spring/Summer 1998 collection. Jakob Schlaepfer, a Swiss company, has supplied the designer with countless lengths of white Guipuere lace to evoke the sensibility of Diego Velázquez in addition to black and burgundy metallic thread-embroidered taffetas that conjure Goya's matadors. Whereas Schlaepfer

preparatory collage for
Autumn/Winter 2001

can afford to offer Lacroix wholesale prices, Liljedahl's textiles, commissioned by exact quantity for each garment, start at £1,000.[2]

It would be negligent to discuss the textiles of Christian Lacroix's collections and omit his fascination for the French fabrics born from the pre-Revolutionary, Empire, and Directoire eras. Lacroix has consistently utilised the white cotton *voiles* and whitework stitches of the early eighteenth century Empire gown, but perhaps more notably, he has championed the silk damasks and brocades, taffetas and satins of the late seventeenth century, all Lesage-embroidered in silk floss and silver and gold *filé* and *frisé* threads.

Acquired by Chanel in 2002, Lesage has been responsible for some of the most iconic embroidered garments in couture history, amongst them Elsa Schiaparelli's beaded dinner jackets and Karl Lagerfeld's famous *trompe l'oeil* jewelled dress for the House of Chanel. Lesage also produced the Persian and African embroideries for Lacroix's 1989 and 1990 collections, and has continued to supply the couturier with the metalwork and silk embroideries for the most recent couture expositions. Lemarie, a featherwork firm as reputable and long-standing as Lesage, has consistently decorated Lacroix's gowns with the multi-hued ostrich plumes made popular by Marie Antoinette.

Christian Lacroix's magpie eye has offered textile designers the opportunity to create luxurious *passementeries*, lavish *devoré* velvets, precious metal decorations, and opulent Rococo-style embroidered and brocaded silks. By fusing the cloths of different eras and regions, Lacroix has not only provided a unique fashion design legacy, but a winning example of the limitless creative possibilities of a diversified textile portfolio.

EDC

opposite
detail
Autumn/Winter 1995
asymmetrical 'stained
glass', 'flowers', and
'medallions' lacquered lace
patchwork top over full
ruffled and coiled 'bishop'
moiré skirt

opposite
detail of Quande Même
Spring/Summer 1989
fringed and embroidered
sky-blue ziberline jacket
over strapless *piqué* dress
printed with Napoleon III
flowers

top
preparatory collage for
haute couture
Spring/Summer 2003

bottom
Autumn/Winter 2002
long ornamental tweed coat
with multi-colour rustic
paisley pattern embroidery
with wool flower appliqués
and ostrich feathers

project**alabama**

Project Alabama is stationed in Florence, Alabama, and contracts over 150 artisans to configure and embellish folksy, earthy garments. Founding designer Natalie Chanin and her New York-based business partner, Enrico Marone-Cinzano, have described their workshops as reinvigorated "quilting bees", and their mission to foster a revival of craft by showcasing local talent supports this description.

The fashion company, which began by utilising recycled cotton jerseys to output faddish printed tee-shirts with topstitched cut-outs of animal and plant figures, has consistently championed one-of-a-kind works that showcase reverse appliqués and elaborate beadwork. Stylised floral patterns are stencilled, or more recently, airbrushed onto jersey and then cut to reveal a contrasting ground fabric. Whether affecting harlequin or sunflower motifs, Alabama has collaborated with graffiti artist Chi Modu and illustrator Robert Ryan, respectively, to produce painterly hues that compliment feminine, whimsical garments.

With the exception of Project Alabama Machine, a two-year-old ready-to-wear line supported by Italian manufacturer Gibo SpA, each Project Alabama garment is hand-sewn from seam to surface and executed in inspired neutral palettes enlivened by rich crimsons and indigos. Feather and chain embroidery stitches are commonly used to emphasise couture craftsmanship. Hand, the company's original line, recently introduced a range of new materials, including denims and techno nylons, which allowed their classic frocks and signature patchwork corsets to metamorphise into more contemporary executions such as evening gowns, hoodies, and separates sets. Though Hand comprises finely crafted garments with high-end price points, Project Alabama Machine is hardly derived from industrial or homogenised production. The garments often feature the same motifs as Hand, and even introduce new Alabama textiles, such as hand-blocked *Japonisme* prints appliquéd with rice paper, which were used in the Spring/Summer 2005 collection.

With regular bulk deliveries of recycled fabrics from the Salvation Army, Project Alabama is a model of reusability and environmental efficiency in the garment industry. Hand-craftsmanship and an investment in the artisanal growth of the Alabama Black Belt region have allowed Alabama an expanding reputation for unique American couture.

EDC

top
Autumn/Winter 2004
hand-embroidered 100%
cotton jersey
photo: Reyez

bottom
Autumn/Winter 2004
beaded cotton jersey with
wool crochet elements
photo: Reyez

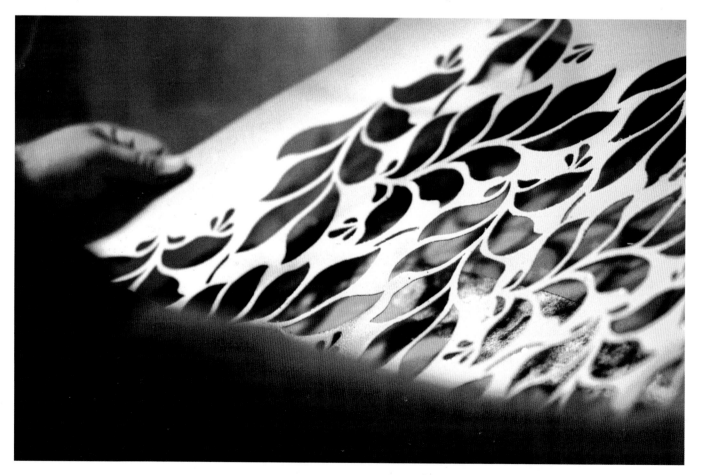

Spring/Summer 2006
cotton jersey
photo: Monica Feudi

opposite top
detail of quilting
Spring/Summer 2006
cotton
photo: Reyez

opposite bottom
stencilling in the Project
Alabama studio, 2005
photo: Robert Rausch

three**as**four

With a collaborative spirit that consistently engages avant-gardism, the ThreeAsFour (formerly "As Four") collective distinguishes itself from New York peers by transforming simple constructive shapes and through layering, embellishing, cutting, or distressing an exquisite range of textiles.

With each passing season, ThreeAsFour's clothing and accessories have become more densely ornamented. While many of the trio's gowns are simple curvilinear forms, emphasising a streamlined, modern silhouette, their most exquisite pieces often feature three or four layers of decorous fabric which reveal their own interior stitch work. The Autumn/Winter 2004 showing featured gowns and tunics of applied leather discs, puckered out to resemble fish scales or tortoiseshell. Executed in metallic and black leathers and outlined in thick trimmings, and overlaying varying densities of fabric, these quirky, sculptural concoctions fluttered and reformed with each step. These siren gowns informed the Spring/Summer 2006 ThreeAsFour Denim line, which juxtaposed 'working man' blue twills and frothy trails of tattered cotton weighted by silver gypsy medallions.

The collaborative's unique layering technique alternately produces fragile *mille-feuilles* or hardened sheaths, depending on the choice of textile. One of their most coveted couture designs is a jacket that the group's members, Gabi, Adi, and Angela fondly refer to as "Cuckoo", both for its schizophrenic application of embossed rayons, hand-painted cotton canvases and silk damasks in hand-cut and glittered puzzle pieces, and for its brilliant configuration, which unsnaps at every seam to form a two-dimensional stylised bird shape.

The spiral is a leitmotif of the ThreeAsFour collection, evidenced by trailing panels of jersey, organza or sheer silk around a bias-fitted form. Like many of ThreeAsFour's illusive, shivering shapes, this form incites continual movement in their silhouettes. The trio has also reinvented the culotte jumpsuit, which appears starkly different when executed in a sheer lame gauze and a white matte vinyl. In fact, ThreeAsFour shows constantly reinvigorate the flirtatious shapes that succeeded in their early lines. The versatility of their work perhaps stems from the distinct aesthetics of the individual designers. The Israeli-born Adi professes a passion for "old movies", and the silhouettes of Jean Harlow and Greta Garbo are channelled in sleeveless gowns with applied ruffles and embroidered tiers of layered tulles, crêpes, and georgettes.[1] Angela's own style serves as muse for the company's glittery, gilded textiles, urged on by a fascination with the Disco-esque wardrobe of American toy icon Barbie. These fabrics were

opposite
Autumn/Winter 2005
silk charmeuse and pleated
silk organza
photo: Caroline Craig Totem

particularly powerful when applied, in polyamide coatings of arabesques and Persian foliate motifs, to Greco-Roman tunics for Spring/Summer 2004, and in lurex patchworks for Autumn/Winter 2005. Gabi's passion for elaborate decoration, particularly in pearl, opal *paillette*, and ribbon beadwork is exploited in custom-made designs, which will soon include embroidered shoes in addition to belts and signature handbags.

ThreeAsFour's creations demonstrate skilled virtuosity through playful and witty embellishments. A ThreeAsFour design, like the collaborative persona of the masterminding trio itself, is an amalgam more technically advanced than it appears.

EDC

embellished

top
Witch Ladder Cascade
Spring/Summer 2006
denim
photo: Marcelo Krasilcic

bottom
Spring/Summer 2004
leather, silk, Swarovski
crystals. linen lurex and
silk lurex
photo: Schohaja

opposite
Autumn/Winter 2005
silk charmeuse
photo: Caroline Craig Totem

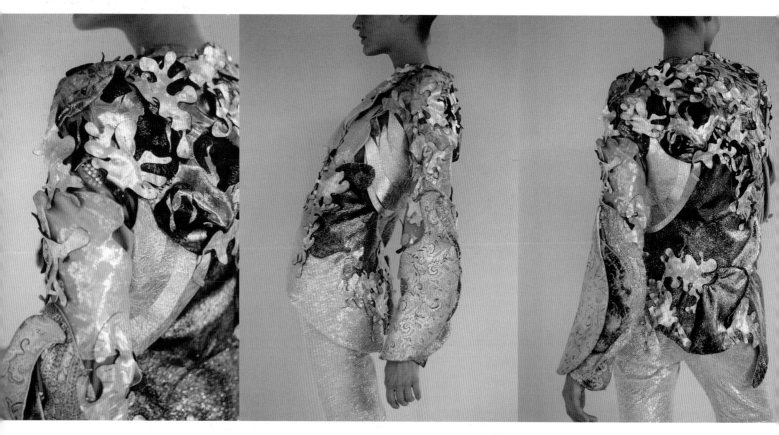

dries**van**noten

Fashion design underwent an irreversible change when Dries Van Noten and his counterparts of the famed 'Antwerp Six' infused the catwalk with wearable avant-garde art 20 years ago. Decades later, and Van Noten has proven himself to be a visionary couturier whose multi-ethnic inspired designs never fail to impress the habitually flippant fashion world. A wanderer when it comes to inspiration, Van Noten has a truly global perspective, reinterpreting existing themes, motifs and traditional hand-crafting techniques from such diverse locales as Northern Africa, India and Japan to create his idiosyncratic designs.

Outré combinations of prints and luxurious textiles, the majority of which Van Noten has designed himself, abound in each collection. Typically natural fabrics including silk, fine georgette, wool, muslin, soft leather and heavy crêpe in a brilliant array of colours can often be found paired with paisley prints, stripes, polka dots and floral patterns, further embellished with beading, embroidery and tapestry. If Van Noten has a recurring theme throughout his collections, it is that he can be perennially relied upon to turn out vanguard, timeless designs that follow his unique vision, rather than succumbing to fashion trends. Each collection draws from a multitude of influences and showcases his penchant

for unconventional colour schemes and print combinations—the defining thread to each garment, however, is an inherent elegance of form.

Autumn/Winter 2007 saw a quietly sumptuous display of autumnal tones in discreet, structured cuts for women—with a hint of Eastern inspiration to cap it off. Notions of India were implied through rich tapestry fabrics and romantic crescent-moon motifs drawn together with thick, regal swathes of bullion needlework. Van Noten proposed this collection as *A Touch of Opulence*, coupling simple, elegant shapes with gilt decoration— sometimes just the merest hint of gold against floral patterns as in a sunflower print on a wool knit sweater, or a sheen of brocade on an austere boxy coat. Van Noten set plain against precious with rich fabrics and intricate gilt embroidery against sober, androgynous shapes—an interesting tension that surfaced to a lesser degree the year before in a womenswear collection denoting "assured femininity and assumed masculinity".

Autumn/Winter 2006 looked to heroines of the 1920s and their implicit femininity for inspiration. Pyjama silks, satins and rustic tweeds made gentle contours of long skirts, cropped and draped trousers, golf pants and kimonos, while delicate embroideries and the inclusion of glass beads, diamonds

opposite
detail of beaded embroidery by Megan Park, who has regularly designed emboideries for Dries Van Noten, produced in India
photo: Sion Parkinson
courtesy Sandy Black

and sequins together with gold threads and metallic yarns added an altogether radiant sheen.

Japan echoed through Van Noten's bohemian Spring/Summer 2006 collection, with Asian influenced delicate fabrics—once again silks, satins and cottons—in kimono style cuts pulled together with obi belts. Various printmaking methods including tie-dye, and botanical motifs and stripes in a gamut of colour made a striking entrance on oversized tailored shirts, low-waisted trousers, slouchy shorts, jackets and skirts belted with ribbons or sashes—a Van Noten signature touch. Hints of metallic brocades on cream through to warm beige tones were enlivened with vivid floral prints in dramatic yellow, amethyst, navy, crimson, orange and bottle green.

Van Noten has inventively, and quite prophetically, melded exotic and Western inspirations throughout his career, winning over fashion critics to his singular vision and aesthetic. With such diversity of style not only throughout his history as a designer, but within each collection, Van Noten's experimental approach and virtuosity as a tailor continues to inspire belief in the art of fashion design.

IC

opposite
Autumn/Winter 2006
discharge printed canvas
embroidered with viscose,
cotton and beads

Autumn/Winter 2006
wool embroidered with silk
and metal yarn

martine**van'**thul

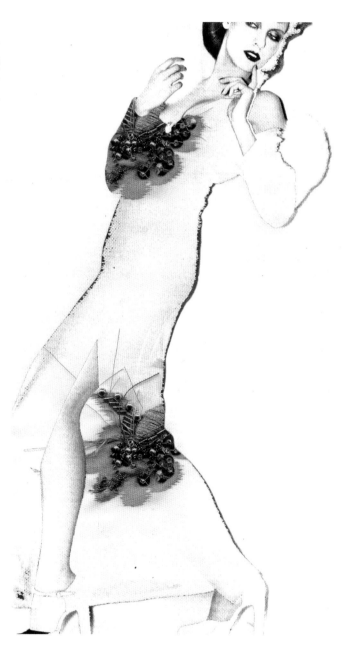

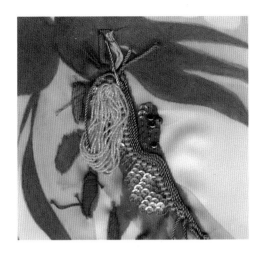

Memory and modernity are held together at once in the work of Martine Van't Hul, a fabric and couture designer whose contrasting use of synthetic and hand-made fabrics evokes a particular notion of time—a time that is both nostalgic and forward-looking, timeless and ephemeral. An expert embroiderer and fabric designer, Van't Hul spent years creating fabrics for Larus Miani in Milan before launching her own collection in 2000, and it is an infatuation with delicately embellished fabrics that has remained at the heart of her work as a designer. The intricacy of the detail in Van't Hul's fabrics, which have included complex layers of fur, feathers, sequins, beads and lace, contain multiple references as she pilfers, pirate-like, from a chocolate-box assortment of periods, layering one upon the other.

The colours that the designer often chooses to work with also evoke a sense of nostalgia: blushing, tea-stained nudes, pearly creams and innocent whites as foundation colours, which are contrasted with splashes of textured appliqué in darker and more intense tones of reds, blacks and browns. The embellishments, which the designer often leaves with hanging threads and fringes, have the look of crushed flowers, found between the pages of a long-forgotten book. This sense of the forgotten and unfinished is a signature carefully woven into Martine Van't Hul's work. Indeed, her first collection was named *Mi-Confectioné*, a term that was introduced at the beginning of the twentieth century, which means "half-produced". Her vision for each garment is that the wearer adds to the design process with their individuality, completing the work that has been started by the designer—a notion that fixes the completion or climax of the garment to a transient moment, conditioned by a specific time, place and wearer.

The idea of the unfinished is evident in Van't Hul's embroidery work in the Spring/Summer 2003 collection, in which thick black outlines of flowers are worked onto cream backgrounds like aged paper, and are only partly filled in with the bright embroidered threads of yellow, red and green that belong to them. Some of the flowers are left uncoloured, like a discarded page of a child's colouring book that have been rescued and glorified for its fragmentary beauty, an innocence that can only be perceived after time has passed. This collection, which was worn by Björk for her world tour in 2003, illustrates the value this designer places on time that is taken, lost and saved.

LMF

detail
Spring/Summer 2005
embroidered silk

opposite left
illustration for an
evening dress
Spring/Summer 2005

opposite right
Spring/Summer 2003
silk and cotton bridal dress
with embroidery
photo: Peter Stigter

sculpted

hussein**chalayan**

Artist, sculptor, showman, and fashion designer, Hussein Chalayan has made the fashion show and performance his consummate medium of expression. During the 1990s the catwalk emerged as a new genre, within which Chalayan demonstrated his mastery of fashion as theatre—not in the overtly spectacular vein of his contemporaries, but in a uniquely studied, thoughtful and intensely cerebral visual experience. By precisely choreographing a range of design collaborators and models Chalayan expresses concepts often rooted in displacement, trans-migration and transformation, making personal, cultural and socio-political statements in addition to creating beautiful objects and fashion for sale.

In Chalayan's hands, the dress becomes emblematic and layered with meaning. He has fashioned dresses from a broad and unexpected range of materials; the 'aeroplane' dresses of moulded fibreglass, with moving panels revealing a froth of tulle; the rigidly constraining carved wooden bodice screwed together with metal fittings; the traditional Turkish costume which gradually transforms and merges with a classic black coat to be reconstituted as a complex modern dress; the chair covers which transform into dresses and the chairs into suitcases, together with the table which becomes a conceptual wooden 'skirt'; the fragile dresses of sugar glass which

are calmly smashed on stage and on film; dresses suspended from helium balloons, or which contain inflatable panels. The silhouette in all these cases is fashioned around a minimalist body, the models behaving as mannequins in the original sense, hangers for the sculptural shapes which the clothes impart.

But although Chalayan professes not to be a textiles person, he develops unique fabrics for his collections including prints, embroideries, knits and weaves, and he also designed knitwear for Tse New York for several seasons. In the *Ambimorphous* collection, Autumn/ Winter 2002, he deconstructs a series of traditional Turkish costumes of vibrant colourful layers of metal embellishments, embroideries and striped woven fabrics, which become jewelled juxtapositions to austere black western dress. *Between*, Spring/Summer 1998, developed a computer graphic print theme, which continued off the body in beaded strands which completed the design in three dimensions. Recent collections have introduced complex and narrative fabrics, representing Chalayan's preoccupations with the circumstances of his native and politically divided Cyprus, showing both contemporary and historical imagery and text, in woven and printed form.

Chalayan is unfailingly referred to as a Modernist; his work addresses the realities of

opposite:
Ambimorphous
Autumn/Winter 2002
traditional Turkish costume
with embroidery combined
with leather and suede
photo: Chris Moore

the present whilst considering the past and looking to the future. He frequently uses clean and streamlined lines with a reductivist minimalist aesthetic, embracing and transposing technologies and integrating notions of travel and place, through deep levels of references. However, the clothes can be understood in fashion terms without knowledge of all these levels of meaning—the integrated use of fabric, asymmetric form and embellishment is sufficiently eloquent.

Series and multiples feature in the work in a number of ways—in *Echoform* a number of similar denim outfits were shown, each one with different stitch and shape details disappearing; the gradual unveiling of the models wearing the *chador* in *Between*; the step by step morphing of garments in stages in *Ambimorphous*, and *Mapreading*, Autumn/ Winter 2001.

Having recently celebrated his first ten years in fashion with a solo exhibition in Holland at the Groninger Museum and the opening of his first shop in Tokyo in 2004, Hussein Chalayan is now a highly influential creative force in contemporary fashion, which is all the richer for his poetic vision.

Sandy Black

opposite and below
After Words
Autumn/Winter 2000
100% cotton dresses/chair
covers and mahogany
skirt/table
photo: Chris Moore

87

husseinchalayan

sanghee**chun**

sculpted

Sanghee Chun's work is the result of her impressive education and industry experience background. Having worked in fashion for over five years at home in South Korea following her first degree, the designer chose to further her studies in the UK at the London College of Fashion. During her studies in London Sanghee was production assistant for both Alexander McQueen and Tata-Naka (a young London-based fashion company) working on pattern-cutting, embroidery, sewing and dyeing. Her blending of print with a range of cutting and construction techniques is what set her MA graduate collection apart and earned her a job at Alberta Ferretti in Italy.

Having always excelled in the manipulation of fabrics, using embroidery, pleating and colour blocking to create three-dimensional pieces, she describes her work as "a combination of art, fashion and technology".[1] Chun's printed and sculpted pieces are highly refined and show a confident understanding of colour, structure and cut. Although her work is heavily pattern-based, Chun has only recently started designing her own prints. Working with a combination of digital print techniques and transfer printing, she is inspired by the idea of creating illusions, such as shadow art and kaleidoscope effects. Chun's graphics blur into one another, in a rich palette of bright purples and a spectrum of green, from olive to jade, accented by tones of cream and pale peach.

The prints from Chun's MA collection, entitled *The Stratagems for the Art of Illusion*, took inspiration from the plants and foliage in Frida Kahlo's paintings. The hand-painted floral design was exploded and intercut with another colourway to match the pleats and create a feeling of constant change and movement. In this way, the prints are disrupted by pleats, layered flaps, frills and structured into overlapping rosettes. The designer's skilled pleating techniques require the use of synthetic fabrics such as polyester satin to achieve the most dramatic effects, with the colour density achieved by duchess silks, georgette and chiffon. Working on the stand, the designer folds and sculpts sections of her garments, drawing particular attention to a segment or panel with the printed fabric spreading out like a concertina, emphasised by every movement the wearer makes.

ER

opposite left and right
The Stratagems for the Art of Illusion, 2006
London College of Fashion graduate collection
silk georgette and polyester satin, hand-painted image manipulated and digitally printed

overleaf
detail
The Stratagems for the Art of Illusion, 2006
London College of Fashion graduate collection
silk georgette, duchess and polyester satin

comme**des**garçons

Both patently shocking and structurally formidable, the innovative clothing of Rei Kawakubo has transformed the ambitions of modern fashion since the inception of her label, Comme des Garçons ("Like the Boys") in 1973. Kawakubo has been instrumental in establishing the language of deconstruction, both physically and theoretically, in late twentieth century tailoring and has similarly catalysed the use of technologically advanced synthetics for higher performance and more hyperbolic dimensionality on the global runway.

The Tokyo-born designer, who graduated from Keio University in 1964 with a degree in Western Art History, actually discovered fashion through its textile building blocks as a promoter at Asahi Kasei, a Japanese chemical company that manufactured the largest number of acrylic fibres in the late 1960s. Kawakubo's attention to technological exploration and production has not waned, and her collections, produced collaboratively with Hiroshi Matsushita and the Orimono Kenkyu Sha Fabric Company, experiment with radical sculpting, fraying, dyeing, and post-weave treatment techniques, as well as the continual creation of new synthetic and natural fibre blends and atypical weaves. Matsushita is also her conduit to over 50 other textile firms who are consistently providing the designer with new fabrics. The cloths of Comme des Garçons prêt-à-porter

womenswear, Tricot (knit), and Homme lines are produced six months to a year before a collection is released, and are realised before patterns are produced, allowing Kawakubo to cultivate and conceive entirely from these raw materials.

The designer's collections require a fusion of traditional Japanese textile treatments, including pre-washed and sun-dried cottons and colourful tie-dyes, and high-tech fibre applications like bonded rubber, polyvinyl-coated cottons and linens, and elastic and rayon blends. The designer also developed new strategies for tactile variation, from scraping natural twills and satins to produce a lacy effect to loop-stitching more threadbare textiles to give them additional girth and weight.[1] Rei Kawakubo rejects the uniformity and industrial efficiency of high-powered loom production, even half-joking at one point that her company must "loosen the screws" of the machine to produce a fabric that is unique in its anomalistic structure and therefore worthy of couture-quality tailoring.[2]

Kawakubo's first blockbuster Paris presentation in 1982, deemed a funereal march for its plenary application of saturated blacks, was unofficially entitled Lace and consisted of boiled wool and cotton knit sweaters, dresses and separates with gaping and frayed holes and slits intermittently

Spring/Summer 2006
natural cotton twill
with white plaster and
metal crown

Spring/Summer 2006
polyester/nylon blend
organza and printed cotton
with hand-decorated beige
paper crown

throughout the pattern; the collection was intended to convey a sense of hand-made opulence, but in fact served as an avant-garde reference to deconstruction and stark monochromatism. *Lace* served to establish Kawakubo as a fashion radical, an image she has only embraced and fostered as the years have passed. Her signature use of black is employed not in the typical Gothic or formal contexts, but to promote her original constructive and woven structures.

Kawakubo's fondness for deconstruction and disjointedness has been evidenced throughout her career, notably with a 1992 collection that featured paper patterns and unfinished clothing, as well as with an Autumn/Winter 2000 runway presentation of synthetic form-fitting bras, cotton velvet boleros and classic rayon negligees. The latter exposition interchanged elements of inner and outerwear textiles and compositions, and reiterated Kawakubo's predisposition to asymmetry. Kawakubo utilised rayon and elastic criss-crossed fibre textiles, pioneered by Hiroshi Matsushita, in 1984 to configure bubbling, gathered, off-kilter cloaks.

Yet for fashion journalists and awed consumers alike, Comme des Garçons's most potent distortions emerged from Kawakubo's "lumps-and-bumps" series for Spring/Summer 1997. The garments, made from nylon/elasticine blend ginghams and polyester/nylon blend chiffons, were tautly stretched and posited as form-fitting dresses and separates. With bulbous insertions placed along the spine, off centre at hip and stomach and projecting from the torso, Kawakubo's silhouettes were both fascinating and horrifyingly disfigured to the Western eye. She reconfigured these structures in Spring/Summer 2006, this time with Union Jack fabrics stuffed inside pouffed sleeves and British and Scottish kilt tartans swathed along the bias at torso and hip lines and ruched into side seams. The effect, despite the historical and Punkish textiles employed, was more self-referential than sociological or historical. By reinterpreting traditional Japanese textile treatments and kimono-based layering to conform to her self-imposed standards of technological and sartorial discovery, Rei Kawakubo has helped to redefine the forums of dress. Her progressive approach to the creation of unconventional clothing through the use of atypical fabrics ensures her continued impact on twenty-first century fashion.

EDC

carla**fernández**

Carla Fernández has a revolutionary approach to clothing, bridging the gap between traditional Mexican communities and the world of high fashion. For Fernández, dress and textiles are the epitome of México's cultural production and she strives to display the vastly rich textile possibilities through her unique design studio Flora 2.

Fernández formulates ethical, fair-trade solutions for the commercial fashion arena by utilising traditional Mexican textile manufacturing and changing the relationship between rural communities and big business. For example, in response to commonplace blue jeans, she creates hand-woven denim using native waist-loom methods: "now if [Mexicans] want, they can wear modern clothes, made by themselves without losing this ancient technique".

Fernández does not work seasonally; instead, she continually designs six parallel collections, devoting each to local communities. She draws inspiration from the signature patterns, shapes and textures of the varying ethnic groups, whilst establishing a network to support hand-made textiles. Native communities also inspire her palette: "At the moment I am enticed by the colour choices of the Chamula women's costume in Chiapas. It's a fine tuned balance of black with very small touches of flashy colours,

even phosphorescent and metallic threads…. I am very interested in the interplay of the textiles and sunlight."

Her experimental, humorous style often focuses on the environment, recycling a variety of found objects from plastic tablecloths to taxi seat covers, and creating coats from blankets and sleeves from rubber tyres. Fernández employs quirky and original techniques such as burning cigarette holes through cotton and dying wool with natural pigments, coffee, and even mud. She freely experiments with garments from past seasons as her work is unaffected by the constant craving for change of the fashion system. Her philosophy is to "go beyond the idea of tradition as something immutable and fashion as something ephemeral".

Zuki Turner

opposite
blanket dress, 2004
hand-woven wool
photo: Diego Pérez

junko**koshino**

Though hesitant to classify herself solely as a garment designer, Junko Koshino has created some of the most unique forms in fashion over the last 30 years. She has introduced cloth sculptures, both Futurist and traditionally-minded that incorporate severe plastics and rubbers as fluidly and masterfully as felted woolens. Koshino's mother, Asako, and two sisters, Hiroko and Michiko are also fashion designers, but Junko has concentrated her efforts the most intensely on pioneering a contemporary aesthetic that focuses pointedly on textile manipulation and tactile exploration.

The crux of Koshino's mission is translating the notion of Taikyoko, or the ability to harmonise conflicting or contrasting elements within a design. While the 1990 *Art Future* Junko Koshino premiere at the Metropolitan Museum of Art provided Koshino the opportunity to showcase structures of gold metal-flaked mesh and sculpted neoprene, allying her with atypical material applications, the 1992 exposition *Taikyoku* delineated the designer's atypical approach to construction. The garments in this exhibition, like those in several of Koshino's consecutive fashion collections, offered playful Japanese textile figuring techniques of creasing, folding, and binding. Executions of *kasaneruto* (to pile), *oru* (to fold), and *tatamu* (to clear) have birthed sensational shapes and surreal forms that toyed with qualities of rigidity and fragility, inflation and collapse—configurations that emphasise the balance of extreme elements.

While Koshino's Autumn/Winter 1986 used an engineered felting process to compress natural fibres of cotton and paper into thick, rigid shell jackets that were both soft to the touch and fortified in their structure, by the early 1990s, she had embraced synthetic materials. Her 1992 creations channelled the architecture of the cube, with boxy quilted plastic coated canvas blocks cinching fitted cotton lycra bodysuits, and the circle, with spiraling and cocooning wires stretched taut with synthetic jersey. At the same time, they also embraced more ambiguous aesthetic concepts, like reflectivity and movement. Koshino revitalised Pierre Cardin's and André Courrèges's 1960s space age ethos through neoprene-covered fibreglass tubing and rubberised curvilinear jumpsuits that strived towards an androgenous Futurism.

The most literal manifestation of *Taikyoku* can be traced to Koshino's brilliant fusion of traditional Japanese imagery and the dialects of Modernism and emerging technologies. She has explained, "I feel that the more modern our lifestyle becomes, the more we feel compelled to revive traditional Japanese elements with a sense of repose."[1] Koshino's collections are the first to elaborate on the whimsical image of the kimono through the

opposite
Cubisme
Autumn/Winter 1991
wool and 'enamel kilted'
nylon cubes

overleaf left
Ibitsu
Spring/Summer 2005
triacetate/ polyester blend
dress and nylon and
wool coat

overleaf right
Neoprene
Autumn/Winter 1992
nylon-bonded neoprene

witty cast of a modern lens. Since her Paris *prêt-a-porter* debut in October 1977, entitled *Primitive Oriental*, Koshino has oft utilised a quintessentially Japanese palette, centered on derivatives of black, white, and nandin, or red bamboo, hues that lend organically to recent trends in minimalism and restraint.

A calculated return to time-honoured Japanese constructive and embellishment methods informed her Autumn/Winter 2006 collection, which fused kimono obi and outer robe wrappings of heavy embossed and brocaded silk with sculpted collars and bustiers of slashed neoprene. Nearly every model wore a cumbersome looped hairpiece superstructed by a moulded plastic and twined thread court headdress, and several were cloaked in open kimono robes.

A master of aerodynamic sculpture, Junko Koshino's fashion designs are always exploratory in their form. She has translated *Taikyoku* through the use of counterintuitive (if ultimately complementary) constructive and textile components. Utilising metallics and mattes, synthetics and natural fibres, traditional weaves and modern technics, Koshino champions a diverse portfolio that seamlessly melds dialects of Futurism and traditionalism.

EDC

ezinma**mbonu**

sculpted

Enthused by the London metropolis that surrounds her, Ezinma Mbonu delicately fuses aspects of urban life throughout her designs to create textural, sensual garments perfectly suited for that same inner-city landscape from which she draws inspiration. Motivated by her curiosity in people, Mbonu sees the inhabitants of London as her muse, "I have always had a voyeuristic tendency, and love observing people in their variety of shape, size, colour and creed. Their varying attitudes to dress is forever an exciting starting point."[1] Like cosmopolitan London, Mbonu's work is a fascinating blend of sculptural forms and textures, with merged cultural references and an overwhelming emphasis on colour. The mesmerising mix of lineage, roots and ethnicity is expressed through a melting pot of textiles, tones and silhouettes. "My style is eclectic, like myself, like London. Eclecticism is never boring or dull, it's always reinventing itself."

Mbonu describes her favourite accessory as colour, for its power to change moods and ability to create visual interest. Lime green stands alongside dusty teal, pale yellow and rusty orange, within delicate, organic prints. Divergent textiles and tailoring mirror the designer's contrasting palette; billowing, feminine silk sleeves complement fitted trousers with industrial fastenings, and plump, oversized knits are paired with neat ruffles and rhythmic pleats. These disparate mixes create edgy, romantic garments—providing a stark contrast to understated urban attire. Exaggerated, divergent shapes are formed as gathered, languid fabrics are puckered in at necklines and draped sinuously in loose layers, balancing against structured, tight sleeves and sharp knife pleats. Mbonu explores a variety of moulded, sculptural forms by contorting, draping and pleating fabrics across bodices.

"It's all about movement, fluidity and shape", Mbonu explains, which justifies her recurring use of the softest, smoothest and lightest fibres such as fine silks and delicate cashmere. Fabrics are fundamental throughout her creations, she is inspired immensely by the individual properties of various textiles and exaggerates their essential qualities to the extreme, utilising the chunkiest woollen knits, relaxed loose jersey and creamy, fluid silk.

ZT

modelling calico toile
on a stand
August 2005

sculpted

top left
watercolour, acrylic and
photoshop collage study for
Autumn/Winter 2003

bottom left
modelling cotton on a stand
August 2005

right
watercolour, acrylic and
photoshop collage study for
Autumn/Winter 2003

left
Autumn/Winter 2003
lambswool and viscose top
with silk georgette skirt

right
Autumn/Winter 2003
lamb leather and cotton

rowan**mersh**

sculpted

Rowan Mersh is a fabric sculptor, a manipulator of extraordinary experimental structures which also relate to the body, immersing it, hiding it, enfolding it. His material is stretchy knitted jersey, together with hardware such as records, CDs, coins, polystyrene balls, cocktail sticks—the list is surprising. Mersh works with these unusual ingredients to bring out the inherent qualities of the knitted fabric in truly innovative ways, which have already gained him the cover of *Crafts* magazine and involvement in a high-profile commission for the set of the Mercury Music Prize. He is a true artisan, a craftsman who creates from experimentation with his materials, the pieces emerging from the relationships between himself and the fabric, the object and the fabric, and his intuitive response to these inspirations.

Although recently graduated from the Royal College of Art's Mixed-Media course in Constructed Textiles, Mersh has amassed an impressive body of work from this and his previous undergraduate degree at Loughborough, well known for its broad approach to textile construction. His work there included exploring new processes, using heat to bond anything from beans to glass to his fabrics to impart a kinetic movement and drape. He also scorched and punched surfaces, distorting woven fabrics through hand manipulation of their warp and weft threads into three-dimensional surfaces—work which attracted the attention of commercial fabric innovators Jacob Schlaepfer.

However, it was the malleability of simple knitted jersey that became the perfect foil for his expression. There is a mathematical precision about the way Mersh works with the cloth, produced industrially, with its regularly spaced stripes providing a base reference point for his distortions: he sews channels into the length of the fabric into which his objects are positioned—or often forced—placing, for example, each CD accurately on the line of a stripe to echo the concentric circles of the object inside. Most of his works are made in one piece, avoiding unnecessary seams, and occasionally using tubes within tubes within tubes, as in the DNA inspired sculpture, which is based on 45 degree pleats along its 16 metre length, incorporating polystyrene balls in decreasing sizes. Mersh's pieces straddle the boundaries between artwork and fashion, craft and performance, commanding attention in their seeming complexity, which is belied by their commonplace materials.

SB

Helix, 2006
striped knit stretch jersey
and approximately 3,000
toothpicks

sculpted

left
80 Metre Skirt, 2006
80 metres of hand-
manipulated striped cotton

top right
Wings, 2006
striped stretch jersey,
polystyrene balls and
barbeque skewers

bottom
Deconstruction/
Reconstruction Part 2, 2006
printed and laddered jersey
with glass beads

Mercury, 2006
silver lycra and wooden discs

isseymiyake

Issey Miyake has always relied on opposing states of stillness and movement within his textiles to create a dialogue between clothing and the underlying body. Commissioning hundreds of textile artists like Koichi Yoshimura and Koji Hamai, Miyake had configured avant-garde wrappings with atypical materials such as rubber, polyurethane resin, and plastic by the mid-1980s. Technologically astute fabrics have accorded the Miyake Design Studio, established in 1970, the freedom to produce dynamic garments that possess a playful physical ambiguity in motion.

Miyake has explained his fascination for the unique capabilities of textiles: "Once it's made, fabric is like the grain in wood, you can't go against it. You know what I like to do sometimes? I like to close my eyes and let the fabric tell me what to do."[1] A strong example of how Miyake's designs make the most of the innate qualities of the textiles is his Seashell coat of Spring/Summer 1985. The dress—composed of nylon threads as fine as fishing line that underscored each bias linen and cotton blend rib—bounced and rolled along the catwalk, mimicking the sea's waves.

Issey Miyake began to manipulate pleated textiles as early as 1988, but his official *Pleats Please* exposition was not until 1993. The 1994 Flying Saucer dress, a slinky full-length sheath of accordion-creased, sun-shaped discs, in multi-coloured polyester, is perhaps the most quintessential of Miyake's pleated creations. These saucers circled the body in a wonderfully disjointed dance, collapsing or expanding with the movement of the wearer. Miyake's unique process for the construction of these garments, usually made from polyester or a blend thereof, gives them a lifelong inscription of pleats.[2] Miyake has worked almost solely with Inoue Pleats Incorporated, Tokyo, to contrive this anatomy of grooved cloths.

Since the establishment of Naoki Takizawa as Creative Director in 2000, a persistence of innovative textile applications has dominated the studio's work, and the resulting garments are both fondly derivative and fantastically divergent from the aesthetic ethos of the house's namesake. Miyake's Waterfall dresses of Autumn/Winter 1984 featured silk jersey drapery at the shoulder and trailing skirts. Whimsical Hellenistic folds were severely countered by an acetate and polyurethane resin torso moulding which, as a hardened breastplate, served to suspend the motion of the piece at its constricted mid-section. Takizawa's Autumn/Winter 2000 translation of the Waterfall Bodies design was manifest in ample fur trim, multi-hued felt and shearling 'breastplates', which provided a static focus for cowling knit dresses and separates. On the

Autumn/Winter 1999
polyester
photo: Anthea Simms

runway, the titillations of fur and shearling fostered a cleverly illusive silhouette, even against the canvas of the fitted base garment.

Naoki Takizawa's work has perpetuated Miyake's career-long focus on atypical forms and ever-evolving shapes through creative applications of high performance textiles. He pioneered jackets and suits of heat-embossed polyester for Miyake: egg carton surfaces whose bubbled peaks could be puckered either in or out. This collection, for Autumn/Winter 2000, also featured voluminous gowns of grey parachuting made hyperbolic with inflatable nozzles that bottled large quantities of air, literally ballooning the fabric away from the hip and torso lines.

Takizawa has certainly subscribed to Miyake's interest in textiles that flutter and quake with an energy of their own, interfacing constantly with the live mannequin beneath them. Yet many of Miyake's most potent creations literally encased the human form, acting as a functional second skin. Takizawa's layered membranes of fluorescent knit scraps connected only with garter clasps, for Autumn/Winter 2000, referenced Issey Miyake's iconic collection of synthetic mesh bodysuits, dubbed *Tattoo Body*. Both presentations, conceived from synthetics that could mimic the regenerative stretch capabilities of the epidermis, generate

a dialogue around the textile's (and the garment's) dual functions of protection and comfort.

Just as both designers have allowed cloth to respond to the body's curves and fluctuations, each has also built structures that exist independently from the skin. Like many of Miyake's pleated products, which extended considerably from the torso in impregnable pyramid protrusions and rolling blisters, Takizawa has pioneered several garments that interact minimally with the model's diminutive frame. Takizawa's Spring/Summer 2006 presentation emerged from a forest of bamboo shoots, and the finale of the presentation was a caged crinoline and bodice of bamboo caning, unlined to reveal the nude body beneath the slats. The bamboo lengths, fitted into a band at hem, sashayed left as the right foot extended, while the immobile bodice served as an effective cage for the upper torso.

Whether executing forms alien to the natural human frame, or creating fluid sheaths that interplay lightly and cleverly with the body, the Issey Miyake portfolio, enlivened each season by Naoki Takizawa, consistently presents the textile as the breeding ground of sartorial invention.

EDC

Flying Saucer dress
Spring/Summer 1994
polyester
photo: Anthea Simms

opposite
Autumn/Winter 2000
cotton jersey, faux fur, wool
blend knit and polyester
ultrasuede
photo: Anthea Simms

Autumn/Winter 2000
polyester blend parachuting
with clear plastic nozzles
photo: Anthea Simms

yohji**yamamoto**

sculpted

Yohji Yamamoto constructs and more often deconstructs clothing in order to challenge contemporary fashion convention. Toying with notions of modern femininity, Yamamoto creates garments with dense areas of pleating and coils of knotted fabrics, rather than conventional ruffles, frills or bows. The simplicity, modernity and androgyny of his creations transform them instantly into timeless classics for both sexes. For Yamamoto, minimalism does not equal uninteresting, and his seemingly simple clothes are given greater depth by the use of complex cutting and construction techniques.

The various approaches Yamamoto utilises throughout his garments strongly reflects his approach to fashion as a whole. Asymmetric shapes and misplaced fabrics parallel Yamamoto's own opposing combinations of simple creations coupled with complicated construction, which can be described as "architecture for the skin". Yamamoto often displays Grecian style pleating, perhaps a nod back to innovators such as Madame Gres, yet takes past techniques into new, untouched realms. Streaks of pleats fall from unexpected areas, beneath shoulder blades or across necklines, elegant twists and rolls replace seams and darts, and fabric is manipulated to create oversized silhouettes and unanticipated shapes. Textiles are fanned into elegant loops and trails or bunched together to create volume and used as a medium to sculpt, rather than to merely decorate.

Yamamoto is strongly influenced by his Japanese heritage, and the geometric forms of indigenous Japanese clothing resonate throughout his work. He uses historical silhouettes such as traditional kimono shapes alongside contemporary streetwear techniques and detailing in his haute couture garments to suggest a postmodern urban feel, with its accompanying connotations of function, protection and durability. This ethos of integrating disparate elements resounds throughout Yamamoto's work. The designer's prolific experimentation with textiles gives his ultra-modern creations a surprising edge by constructing them from natural materials ranging from felts and cottons to muslins, and even stretches as far as using wooden panelling secured with bolts and hinges.

Another element of Yamamoto's heritage is displayed through his disregard for the constantly changing, evolutionary nature of fashion. He sets his own pace, where shapes, silhouettes and colours come from his own personal influences, rather than the latest trend. "People of my generation were ripped off by economy, during our youth, the industry kept pumping out new products we couldn't believe in, because we knew, come tomorrow, they would be out of style."[1] This disregard for

opposite
Spring/Summer 2006
cotton

the industry's speed is clearly displayed in Yamamoto's work, as he persistently refines his palette of neutral colours, which perfectly suit his sexless garments. Sombre hues of black, navy and white prevail throughout his collections, with hints and occasional flashes of colour for impact, rather than a rainbow of hues that adapt to conform with fashion's up-to-the-minute standards.

The timeless nature of Yamamoto's creations is played upon by the designer, and he believes that time can only add to the beauty of his creations. "My clothes are about human beings: They are alive."[2] This statement was tested when Yamamoto chose to display some of his garments in a garden at the Hara Museum in Tokyo in 2003. There, his creations were subjected to a pace set by nature rather than fashion, and were laid bare to rain, sun and wind, ultimately for the textiles and textures to be improved by the natural wear of time.

ZT

Spring/Summer 2006
cotton

Spring/Summer 2006
cotton

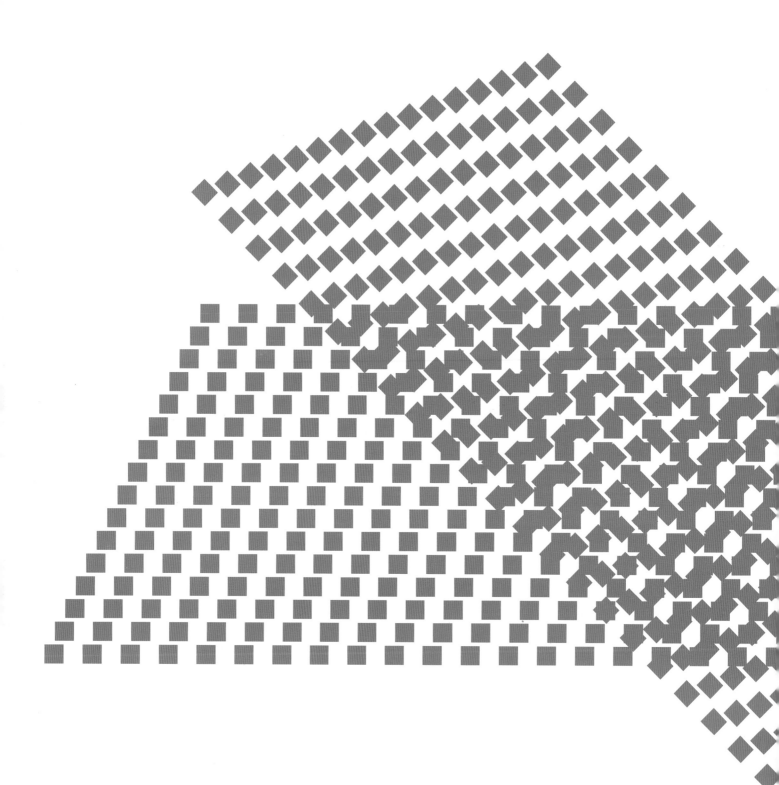

imprinted

basso&brooke

imprinted

opposite left
Vanity Affair
Spring/Summer 2006
digitally printed silk
organza/silk satin corset
and skirt with organza
rose bouquet hat by
Stephen Jones
photo: Chris Moore

opposite right
Vanity Affair
Spring/Summer 2006
digitally printed quilted
silk satin with hat by
Stephen Jones
photo: Chris Moore

left
Vanity Affair
Spring/Summer 2006
digitally printed silk satin
photo: Chris Moore

right
Succubus and Other Tales
Autumn/Winter 2005
digitally printed cotton
poplin shirt, wool suiting
waistcoat and leather
gloves with suede mask by
Stephen Jones
photo: Chris Moore

Constantly redefining the scope of the fashionable print, Bruno Basso and Christopher Brooke employ an eclectic spectrum of couture imagery, which is absorbed into effusive illustrations that challenge, despite their mechanised production, the very definition of the haute textile. Basso & Brooke pioneered a reputation for subversion by exhibiting stylised male and female genitalia as composites of a larger, more aesthetically refined print for their first collection in late 2004. Entitled *The Garden of Earthly Delights*, the presentation championed their patented digital printing technique, which has the ability to saturate a broad range of fabrics from rough leathers to silk and wool jerseys. Having manipulated *ombré* shades of navy, sunflower, and crimson in painterly strokes, Basso & Brooke are inspired colourists, and their engineered hues are enlivened by an allover application to the ensemble from glove to stocking.

The most captivating element of the Basso & Brooke runway spectacle, eclipsing even audacious thematic arrangements such as *The Myth of Succubus* (a medieval legend of a she-demon who seduces men in dreams), is the collaborative's ability to revisit the shining triumphs of haute couture past by printing details that were formerly executed through cut or embellishment. The prints often speak to exaggerated interpretations of femininity, from demure to outright formidable. Basso

& Brooke's Spring/Summer 2006 collection honoured "disenchanted housewives... for whom bridge parties and coffee mornings have been superceded by pills and pitchers of cocktails before noon". This warped kitsch was conveyed masterfully by a revival of the nubby hounds-tooth check, printed on wasp-waisted cocktail dresses and sweetheart ballgowns as a reference to 1950s cocktail couturier Christian Dior's fondness for the enigmatic weave. The signature Chanel bow, which appeared in seamless bugle bead embroideries or inset appliqués on Gabrielle's 1930s gowns, was also revived, but on a sheer silk sundress repeat-printed, complete with trailing ties, to emote a sugary, submissive quality. In their Autumn/Winter 2005 presentation the designers recalled the sculptures and vases of the Greco-Roman era as seen through the lens of Gianni Versace's early 1990s interpretations of the same works. The House of Versace made the golden chain (often with a Medusa pendant) a recurring symbol of the company, and accessorised eggplant and red silks with belts, buttons and necklaces of these links. Basso & Brooke smartly printed hyperbolically layered strands of the same interlocking chain against a sensual palette of deep purples and dusky scarletts.

Basso & Brooke prints have reconfigured the mythical tresses of Rapunzel into a digital collage worthy of the Klimt canvas and reinvigorated Toulouse Lautrec's notorious red-lipped can-can girl into a cartoony contemporary diva. The pair's bustle-back dresses and leg of mutton sleeves, while cleverly executed with the assistance of multi-national conglomerate AEFFE's financial backing, are hardly visible as the backdrops to an astounding portfolio of complex printwork.

EDC

Succubus and Other Tales
Autumn/Winter 2005
digitally printed silk
crêpe-de-chine
photo: Chris Moore

overleaf
detail of Vereda Tropical
print

eley**kishimoto**

Throughout a career of commissions for such prominent fashion firms as Alexander McQueen, Jil Sander, Givenchy, Yves Saint Laurent, and Louis Vuitton, design duo Mark Eley and Wakako Kishimoto have cultivated a rich portfolio of graphic prints that draw equally from historical and modern references. Their Autumn/Winter 2005 collection, *Fairy Tales and Fine Tailoring* took a romantic turn with Empire dresses and sheer sheaths printed with the locks of a fairy tale princess, murky forest brush, fractured teardrops, and flames, while the confections of the following season's *Cosmic Dolls on Earth* showed printed lips, cherries, unicorns, and hearts on Broderie Anglaise sundresses.

The company also produced innovative fabrics for both Hussein Chalayan and Joe Casely-Hayford in the mid-1990s, including a pixellated hand-drawn floral motif for the former, and a Sun print for the latter. Perhaps Eley Kishimoto's ability to translate the medium of print to the contemporary textile prompted high-end sportswear company Ellesse to appoint them as their joint creative force. Eley Kishimoto revived and reconfigured the Ellesse penguin, a company logo since the 1970s, for the Autumn/Winter 2005 collection, using playful repeat prints of mountainous slopes, wintry forest scenes and stylised penguins on neoprene (a synthetic rubber) fleece. Yet for all of their technological inclinations,

Eley and Kishimoto are equally well versed in traditional motifs. In Autumn/Winter 2004 they exhibited an array of daedal patterns composed of lush bunches of grapes and brightly coloured bamboo shoots configured into ovals. The painterly quality was certainly reminiscent of Raoul Dufy's *Toiles de Tournon* block prints, yet many of the compositions derived from the agricultural-themed Soviet Constructivist fabrics of the same era. With a similarly historicist eye, Eley Kishimoto produced an aptly-named Damask print for an A-line coat featured in their Spring/ Summer 2003 collection, clearly inspired by Renaissance-era Venetian *Scarlati* (diagonally woven twill-effect wools whose name derives from a deep red grain dye). A composition dominated by the iconic pomegranate motif was shaded to affect the tonal range of voided velvet. Conversely, a range of bright, psychedelic Op Art prints for Spring/Summer 2005, and a sharply tailored three-piece suit with an all-over Bauhaus colour block print for Autumn/Winter 2002, drew upon more modern precedents.

Despite this breadth of influences, Eley Kishimoto is consistently loyal to the simple, graphic repeat, a benchmark of the "Good Design" era of the mid-twentieth century. Good Design marked the heyday of hand screen-printing in America and Britain, and its proponents included Chicago-based Angelo Testa and British designer Lucienne Day, amongst others. Testa's most celebrated works comprise cartoonish two-tone prints with jagged, naïve compositions. His clean-cut lines, urban tones, and powerful use of colour had a marked influence on the Spring/Summer 2005 Eley Kishimoto collection, which abstracted the buildings of the couple's Brixton neighbourhood in campy, casual prints.

Eley Kishimoto has channelled Lucienne Day as well, whose successful 1951 pattern, Calyx, depicted alternately-sized lampshade shapes in a number of colour ways. Whilst EK's textiles are usually more contained and geometric, they do champion the playful, electric quality that Day helped popularise in the print milieu of the 1950s.

Leaders in the contemporary revival of British printmaking, Mark Eley and Wakako Kishimoto are following in the footsteps of innovative twentieth century printmakers, such as Zandra Rhodes and Celia Birtwell. This British-Japanese design team's rare amalgam of postmodern fusion and contemporary intuition offer Eley Kishimoto's clientele a witty and informed range of garments.

EDC

Cosmic Dolls on Earth
Spring/Summer 2006
screen printed georgette

hamish**morrow**

left
detail
Digital Echoes
Spring/Summer 2004
digitally printed by
Philip Delamore
photo: Sion Parkinson

right
Digital Echoes
Spring/Summer 2004
Virtual print—digital
projection onto white dress
photo: Chris Moore

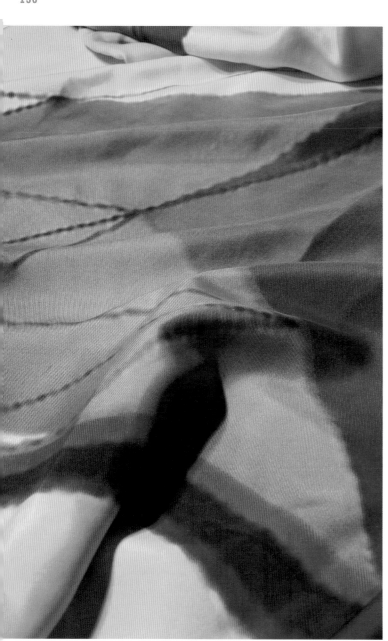

Hamish Morrow's collection from Spring/ Summer 2004 glimmers in dark light, with the glitter and fizz of digital images that threaten to dissipate before our eyes. A South African-born designer who describes himself as "ideas-based", Morrow has partaken in several collaborative works, embracing cutting edge fabric technologies and fine art practices as well as the digital media which he employed for this breathtaking show.[1]

This collection, which was inspired by yachting, features an abstract print in watery ocean hues, recalling overlapping spinnaker sails, on a cloud-coloured background. The dresses are belted and held by haute couture yachting ropes with steel eyelets, which act like a skeletal structure following the line of the shoulder blades, from which the pale sails of fabric hang. The nautical hardware thus weighs down delicate billows that might otherwise have fluttered away in a sea breeze.

Morrow nailed his futuristic, avant-garde colours to the mast by producing the entire collection without the use of a single stitch, employing a process of ultrasonic welding to fix sections of the garment together. The designer collaborated with photographers and image-makers Warren du Preez and Nick Thornton-Jones and multi-media artists UVA for this show, in order, as Morrow puts it, to push "the idea of digital print further" by bringing it into

three dimensions. Models walking down the catwalk in Morrow's designs were filmed through digital filters that abstracted their movements into shapes that mimicked the geometric print on his dresses. The models were thus transformed into the print that they were wearing in one self-reflexive movement—enlivening the digital shapes and breathing life into them. The film was then projected back onto a plain backdrop creating the impression that the models were, in fact, fragmentary extensions of Morrow's print, splitting and dissolving into twinkling forms.

The show ended with a groundbreaking vision for print, as three models appeared in plain white dresses like blank canvases, onto which the luminous print that had been created by filming the models themselves was projected. This fusion of precise digital data and the organic movement of the fabrics draped on the models' bodies created a momentary vision of the future, in which print becomes light and ephemeral, constantly moving, changing and evolving. This fleeting image is, for Hamish Morrow, at the heart of glamour—a 'dream' that fades quickly into digital dust.

LMF

pucci

Emilio Pucci made a lasting impact on fashion with his first ready-to-wear presentation for Lord & Taylor in 1948, supported by the charismatic Diana Vreeland, editor of *Harper's Bazaar*. Yet his kaleidoscopic prints, sportswear, and fluid resort garments found their most loyal patronage from American icons like Marilyn Monroe, in the late 1950s, and Jacqueline Kennedy in the 1960s, when complicated optical patterns and liberated, wearable women's fashions began to govern the global mode. Pucci fanatics gladly discarded the muted tones and prim cuts of the post-war era for his saturated colourwheel and revolutionary 'comfort' constructions, which helped promote the popularity of the resort wardrobe and ultimately bolstered the success of Italian style in the global market.

Officially the Marchese di Barsento a Cavallo, Pucci's aristocratic lineage, which can be traced back to thirteenth century Florence, allowed him a privileged perspective in the creation of fashionable jet-set garments. However, his unique vocabulary of colourful, planar printwork distinguished him as a twentieth century innovator. Pucci's electric range of hues, including azure, tangerine, plum, fuchsia, lime, turquoise, tortoiseshell, pistachio, coral, and absinthe, and his draped gowns, swimsuits, and palazzo and Capri pants ensembles have been consistent market successes throughout the latter half of the twentieth century.

Pucci understood the evolving nature of fashion marketing, and quickly licensed his fabrics, his strongest signature, to upholstery and accessory firms. He designed uniforms and interior textiles for Braniff Airlines throughout the late 1960s, as well as for Qantas, the Australian airline, in 1974. One of his most successful furnishing textiles was the 1963 Fantasiosa print, a primitivist patterning of sunflower yellows and cornflower blues, offset by alarming cherry red accents.

The designer opened his first boutique, La Canzone del Mare (the Song of the Sea) at Marina Piccola in Capri, and utilised Caprese artisans to produce his first handcrafted textiles. Pucci's interest in the development of new synthetic fibres and weaves prompted his collaboration with both Guido Ravisi, a silk industrialist, and Legler and Valle Susa, who assisted him by researching various cotton and silk blends for lustre, drape, and durability. By 1950, Pucci's work utilised mostly synthetic, often nylon-based textiles. As these new synthetics were crumple-free and convenient, resilient and modern, they were the perfect canvas for his contemporary printwork. Jersey is most commonly identified with Pucci's slinky gowns and lightweight beach pants, and with the production of silk jersey by the Mabu workshop in Solbiate, Pucci was the first designer to patent its use.[1] As Pucci's garments require a certain attentiveness

Spring/Summer 2002
polyester blend jersey with
modified Mirror print

Autumn/Winter 2002
pinwheel-printed silk velvet

on the part of his clientele to maintain a body-conscious silhouette, the designer was contracted in 1960 by Formfit Rogers, a Chicago-based manufacturer, to fabricate the "Viva Panty". A stretchy silk and synthetic blend bodysuit, the Viva Panty provided the less-than-perfect wearer with a seamless, nude second skin.

Emilioform, a silk shantung and nylon blend that became the choice fabric for Pucci's early 1960s collections, was also issued in 1960. Pucci's first successes were ski garments, and with the launch of Emilioform, the artist produced a new ski suit, entitled Capsula, which clung attractively to the human form. The Emilioform garments were also distinctive for their translation of a single printed design to all components of the ensemble, from bikini to beach hat to blouse; their complex planar juxtapositions offered an avant-garde camouflage effect. Emilioform-style prints were nostalgically revived for the mini-dresses and raincoats of the Autumn/Winter 2002 Pucci presentation, demonstrating the modernity of Emilio Pucci's vision. This exhibition also revived the designer's frequent use of cotton velvet, which provided a laissez-faire tone to Pucci's beach clothes, shoes, and handbags. The Stained Glass formation of 1965 has proved one of Pucci's most collectable handbag prints, and was executed solely on cotton velvet. Although he had his

preferred fabrics, his prints continue to accommodate nearly any textile surface, from silk and cotton plain weaves to terrycloths, plastics, and paper.

By the early 1970s, Pucci prints could be found at Neiman Marcus, Saks Fifth Avenue, I Magnin, Bergdorf Goodman and, of course, Pucci boutiques worldwide. Pucci was acquired by luxury conglomerate LVMH in 2000, and has recently benefited from creative director Christian Lacroix's reinvigoration of classic prints like Moiré and Vivara (first created by Emilio Pucci in 1966). Even after the founding designer's death in 1992, each House of Pucci textile was cleverly branded with Emilio calligraphy, which originally served to protect the elite status of the creator's name, but has ultimately authenticated the iconic Pucci print amidst its countless imitators.

EDC

zandra**rhodes**

Since her graduation from the Royal College of Art in 1964, Zandra Rhodes has created a complex textile language rife with ethnographic imagery, illustration, and romantic nuance. A pioneer of contemporary fabric design for fashion, Rhodes' passion for the relationship between the hand-screen print and the three-dimensional garment has influenced countless designers, from Jonathan Saunders to John Galliano. Rhodes' colourful vestments are cut to emphasise the textile rather than promote innovative or revolutionary forms, which distinguished her from both the revered tailors and socio-aesthetic deviants of London's 'swinging' 1960s fashion panorama. In *Zandra Rhodes: A Lifelong Love Affair with Textiles*, Brenda Polan remarks, "[Zandra] shifted the world's perception of British fashion from either classically conservative or 'outrageous' to ground-breakingly creative and gloriously baroque."[1]

Zandra Rhodes' RCA degree show print, Medals, which she claims was inspired by David Hockney's *Generals* painting and the revolutionary landscape of Pop Art, was generated coincidentally with her studies at the War Museum, and was eagerly purchased by HEAL's, the reputable British furnishings firm. Dominoes, Explosions, and the first of her Lightbulbs series were all commissioned by Marion Foale and Sally Tuffin before Rhodes opened her own clothes shop on Fulham Road in London in 1969.

Since the commencement of her career, Rhodes' sketchbook has been a worthy collaborator, allowing her to jot down provocative structures, landscapes, or artworks before transforming the renderings into components of a larger print. Though her mode was undoubtedly impacted by fashion's reappropriation of indigenous dress styles and all things 'ethnic', and was derived in part from the explorations of Emilio Pucci as a colourist, Zandra Rhodes provided a new niche of artisanal clothing marked by distinctive print compositions and couture-quality finishing. Her early handiwork, drafted onto printed chiffons, organzas, and broadcloth silks, featured hand-rolled edges and applied trimmings. Her technical ability in manipulating the screen print, as well as her signature vocabulary of exoticism and nostalgia made her clothing a favourite of British *Vogue* editor Beatrix Miller, who in turn helped to popularise Rhodes' designs in New York via then-*Vogue* editor Diana Vreeland. Early proponents of Rhodes' lines were Fortnum & Mason in London and Henri Bendel in New York. With their high-end reputations and upper class clientele, these stores perpetuated the exclusivity of Rhodes' clothing, and helped promote her work as wearable art.

Zandra Rhodes in her London studio, January 2005

left
Waistcoat, 1970
Chevron Shawl print on silk
with an ethnic inspired
quilted silk yoke

right
Kaftan, 1970
silk chiffon printed with
Indian Feather Sunspray

Rhodes' assistant working
on Kodatraces on a light box
for a three-colour design
called Laces Roses

By 1979, Rhodes was preparing one kodatrace and one screen for each colour within her two- to five-toned prints. With lengths of screen printed fabric, Zandra conceptualised her blueprints by draping the textiles on her own body, both for a sense of how they would translate in movement and to witness their composition when set against a human form.

Few visual leitmotifs have been as prominent in Rhodes' printwork as those extracted from Asian and Native American textiles. A squiggly line and a distinctive Z (for Zandra) have composed a recurring calligraphy in the ground of many of her creations, and she has looked to traditional textile crafts to enliven her print formations; her Knitted Circle design of 1968 and the Chevron Shawl pattern of 1970 (which was informed by Victorian embroidery and fringe) both transformed stitches into two-dimensional constellations. Yet a 1970 trip to New York City's Museum of Natural History spurred Rhodes to create her Indian Feather Sunspray series of the same year, which comprised consecutive rows of stylised feathers and was cut into a variety of halter sheath dresses and tunics. Indian Feather Sunspray print offered colours native to the American landscape—most notably gradated terracottas, peaches, and turquoises—and was reinvigorated for several seasons. Rhodes mused, "I wanted to give the impression in the print that the feathers were sewn or embroidered onto the fabric with cross-stitch.... Thus from real feathers I went to printed feathers and from there to cutting the print to make feathered edges.... I did all the cutting out round every feather myself then hand-rolled the edges."[2]

Zandra Rhodes' 1980 and 1987 trips to India, the former as a participant in the *Festival of India*, and the latter to present her collections in Mumbai and Delhi, inspired her use of traditional sari crests and edifices. Alongside her typical silk chiffons, organzas, and georgettes, the artist used *kanchipuram* silks and *Vekatgiri* cottons and dyed them jasmine, saffron, and persimmon tones.

2003 marked the grand opening of the Fashion and Textile Museum in London, which Rhodes co-founded in order to recognise and promote the textile designers of contemporary fashion. With over 40 years of creation behind her, Zandra Rhodes still enthuses about the discipline of integrating fabric and figuration.

EDC

Net Flower Turn-around
imprint on the table backing
cloth after the silk chiffon
fabric has been printed
photo: Sion Parkinson

jonathan**saunders**

Scottish-born Jonathan Saunders has developed a clear handwriting in both fashion and textiles; his distinct signature in printwork is akin to Emilio Pucci's, yet his complex prints also work intuitively within the medium of constructed fashion. From his debut collection in 2003 on, Saunders has been lauded for his bold colour ways and graphic compositions. While Saunders' portfolio is influenced by the pixelated, intricate markings of advanced computerised printing, and his work may convey the frenetic movement and fragmented tonality of the digital age, he works almost solely with hand-screening techniques.

Both rigorously engineered and playfully avant-garde, Saunders' prints communicate movement and vitality, and his designs recall various twentieth century art historical eras. His Spring/Summer 2005 presentation offered printed textiles with Bauhaus linearity and palettes that recalled the weavings of Anni Albers and the colour block paintings of Paul Klee. As his work has evolved, Saunders' prints have further examined colour through Futurist and Cubist planes. These trapezoids and overlapping arcs indicate a fondness for the constructs of Art Moderne, yet Saunders has also cited the earthy tones and raw figurations of sculptor Anthony Caro as an inspiration.

The overwrought composition of Saunders' prints is often countered by the elegant simplicity of the designer's cotton, wool, silk, and synthetic jersey fabrics. In recent seasons, he has also favoured silk and wool crepes and satins, which in his Autumn/Winter 2006 exposition lent a more formal tone. Inspired by Peggy Guggenheim's 1940s art collection, the prints retained Saunders' signature modernity while providing a more comprehensive range of shapes and applications in the form of ombré colour bars along hems, and stark slate and silver coatings, which crackle as they move around the model's torso.

Borrowing a technique pioneered by Zandra Rhodes, Saunders manipulates each pattern to perform to its full potential by matching his painterly prints with the shape of the garment and its seams. His exotic Autumn/Winter 2005 designs applied flecked and fibre markings to knit silks and crepes, mimicking the natural fur or hair traditionally used in African and Asian indigenous dress. One printed synthetic was covered head to toe in scoring that evoked tribal scarification. These embellishments correlated impeccably with the seams of tailored suits. This ability to match printed designs to three-dimensional forms, alongside his fantastical manipulations of graphic patterns, has distinguished Saunders from his peers in the twenty-first century print revival current in London fashion.

EDC

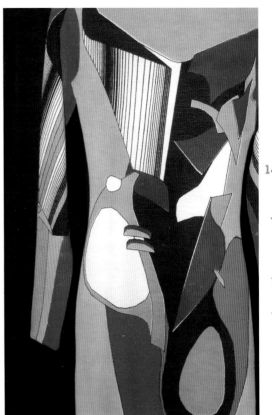

left
Spring/Summer 2006
hand-dyed black, grey and
white chiffon woven
together in basket weave
photo: Sion Parkinson

right
detail
Autumn/Winter 2004
hand-printed jersey
photo: Sion Parkinson.

top
detail
Spring/Summer 2006
hand-printed multi-
coloured and striped
georgette rouleaux loops
photo: Sion Parkinson

bottom
Spring/Summer 2006
pleated and hand-printed
white silk chiffon with silver
foil and black ink
photo: Sion Parkinson

opposite
Spring/Summer 2006
hand-printed silk
photo: Chris Moore

renato**termenini**

imprinted

Predominantly designing with silk, Renato Termenini employs laser-cutting technology alongside ancient dying techniques to create sensuous, stand-out garments with bold prints in vivid colour. Termenini's inventive silhouettes are a nod in the direction of the avant-garde, while an Eastern influence is made plain through the delicate hand-dyed patterning on organic fabrics consistent throughout his collections. As a reaction against the ubiquitous digital print designs of contemporary mass-produced clothing, Termenini has chosen to utilise traditional hand-dying methods to ensure the singularity of each garment. With considerable aplomb, Termenini has taken the oft-derided tie-dye motif to the catwalk—transforming it from characteristic imaging on hippy threads and thrift store finds to couture detailing.

Termenini's physical and artistic approach to fashion is apparent in his striking print designs— the product of labour-intensive dying methods derived from the ancient art of Japanese *shibori* or tie-dye. *Shibori*, meaning to 'squeeze' or 'wring', is similar in a practical sense to tie-dying, but emphasises the necessity of harmony and balance to be struck between dyer and materials. This relates to the interplay between the uncontrollable, and unexpected nature of tie-dying, and the dyer's ability to allow chance and accident to positively contribute to the appeal of the design.

Termenini's collection of *shibori* silk dresses showcase a smattering of almost iridescent fuchsias, silvery greys and tawny yellows across elegant shapes that fall naturally, informed by the curves of the wearer. *Genus*, a more recent collection, saw the distinctive tie-dye patterning take on the appearance of exotic butterfly wings with staccato markings in bold colour bursting across sleek dresses. Furthering this effect, spiral shapes were laser-cut into the fabric to create flowing movement reminiscent of fluttering insects, with the laser's precise finish allowing the garment weightlessness and ease of movement. "I am often inspired by nature", explains Termenini of his influences. "I love to recreate its colours and textures. In particular I love opening a piece of fabric to discover the unique, colourful patterns that have formed inside."

In stark contrast to the silk dresses, Termenini combines a diverse and surprising range of materials within his collections including 'gaffa' tape—in the form of hazard tape trousers—PVC and leather. The designer says, "I like to combine a variety of fabrics with simple shapes, tailoring is a very old craft that is dying. Combining printmaking and tailoring can result in exceptional designs—the possibilities are endless."

IC

Genus
Spring/Summer 2005
tie-dyed silk

carolewaller

imprinted

From her sleek artist's studio and exhibition space in Bath, Carole Waller produces her unique womenswear collections, transforming her hand painted canvases into pieces of 'wearable art'.

Having studied Painting at Canterbury College of Art and Fine Art Textiles in an MA at Cranbrook Academy of Art, Michigan, Waller began experimenting with painting on clothing, seeing it as an effective way of ensuring her work was seen by a wider audience. For over ten years, she has exhibited at the Paris and New York fashion weeks, combining hand-painted printing techniques with cutting edge clothing design.

Using brightly coloured dyes, thickened with a seaweed-based gum, the artist sweeps bold colours across the canvas, adding occasional placement prints of elusive, shadowy figures in the background. While each canvas is hand-painted, Waller had previously looked at ways in which to mass-produce her work, in contrast to her energetic, spontaneous creations, the results seemed flat and lifeless. Now she relies solely on her own hand and eye, choosing colour, print and pattern as she deems appropriate, maintaining the exclusivity of every design.

Comprised mainly of abstract drawings, Waller's prints focus on the human presence in space and time. Illustrated by confident sweeps of colour, a multitude of print and pattern bring energy to the various fabrics, ranging from delicate silks, to crisp cottons, linen and viscose rayon. Waller works with womenswear designer Ray Harris, creating designs which are sympathetic to her prints and chosen techniques.

Her collections combine varying styles of garments, from long bias cut skirts and dresses, to tailored linen blazers, and thick velvet jackets. Inspired by a single image or technique, Waller bases each collection around one theme, creating variations of her original piece to build her range. Methods Waller has used include appliqué, patchwork, felting and needle punching, a technique combining wool and silk to integrate two prints. The artist works within the capabilities and strengths of her chosen fabrics, emphasising the mark-making and dyeing processes in order that they become an integral and deliberate feature on every garment.

ER

Beach
Spring/Summer 2006
dye-painted silk

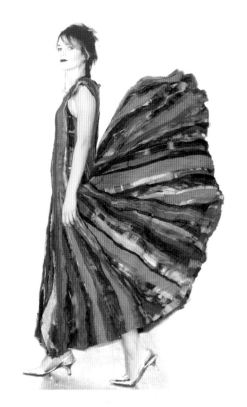

carolewaller

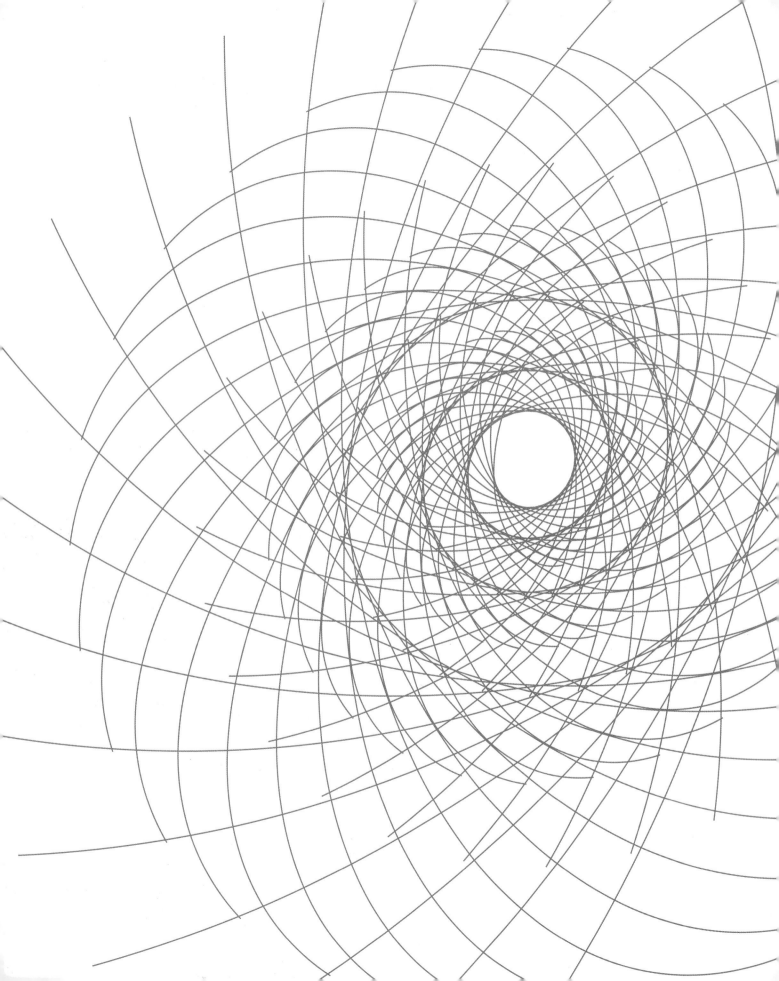

constructed

shoto**banerji**

constructed

Shoto Banerji has been persistent in the creation of rich, opulent fabrics that combine a love of tradition with a passion for the contemporary. Remaining true to age-old conventions, she has formed a body of work, which, inspired by nature and the environment, stands out as both experimental and rooted in history. Working closely with the Maheswari School of handloom weavers in Central India, Banerji has honed her skills as a textile designer, creating fabrics that blend convention with the demands of the modern, international marketplace. Diverse in character, the weavers' practice is strongly influenced by the culture, architecture and features particular to their region. Playing upon and developing their skills as craftspeople, Banerji contemporises the final garment, merging traditional techniques and patterns with new fabrics and a fresh, enticing colour palette of neutral, earthy hues as well as deep crimsons, greens and burnt oranges. Incorporating materials such as Thai silk, Belgian linen and Italian wool, she has updated the process of weaving, giving her fabrics a sophisticated aesthetic appeal, peppered with fine, delicate detail.

Banerji's work frequently evokes her natural surroundings, referencing, for example, the feathers and plumage of native birds or the flowering patterns of the Thai orchid. Translating the patterns and textures found in nature into woven textiles, she creates garments diverse in colour, content and surface. Creating materials that shimmer and transform with movement, Banerji has consistently pushed the limits of design, breaking away from conventional placements of pattern detail and redefining the handloom process from ethnic to exquisite.

Exemplifying this notion of design development, her earlier collections, which combine textiles and ceramics, demonstrate the importance of the fabric in determining fashion. Driven by an ongoing desire to push textile production in new directions, she combined yarns, sourced from Pitti Filati, Florence, for instance, with clay, which were then subjected to high temperatures in the kiln. In keeping with the notion of clay forms, the silk garments are then fashioned on the knitting machine, retaining their shape when worn.

Shoto Banerji has established herself as an important figure in bridging the gap between traditional Indian handloom and hand-woven wearable textiles. As an artist and designer, she has helped to conserve conventional practices while remaining true to the ever-changing, fast paced industry in which she works.

Beccy Clarke

opposite top left
Magpie, 2005
woven silk
photo: Sion Parkinson

opposite top right
Dragon Fly, 2005
woven silk
photo: Sion Parkinson

opposite bottom
detail
Pheasant, 2005
silk
photo: Jean-Pierre van den Waeyenberg

liz**collins**

constructed

Liz Collins' broad experimentation with fabric, including crocheted lace, leather, fur, and high-tech materials, and with methods of treating fabric, such as printing, knit-grafting, and strategic shrinking, is the basis of her trademark style. At the Rhode Island School of Design, where Collins earned both her Bachelors and Masters degrees in textiles, she became highly skilled in machine knitting, discovering 'knit-grafting', the technique by which she integrates a plethora of fabrics into her constructed textiles. Her clothes are created using techniques that allow the fabric and the garment to be realised simultaneously and organically. For Collins, constructing a garment is a concurrent process of creating, embellishing, shaping, and layering the fabric.

Collins says that fabric itself is a "perpetual ongoing inspiration" for her, and her designs reflect an understanding of its versatility and potential.[1] She can be inspired simply by a material, or by the ways in which two materials can be fused together. Her clever juxtapositioning of contrasting textures and shapes can be seen in many of her designs where the flowing layers, made up of bands of fabric, hang loosely from fitted knit textiles, which cling to the body's curves, creating a sense of rhythm. Collins' inspiration, however, goes beyond fabric, and ranges from anatomy and the Gothic to music. The Led Zeppelin fuelled collection of 2003, for example,

featured deerskin, shearling, and a pallette of earthy browns and reds. Consumed, then, by the rock vibe, she created striking garments that evoke the savage. In another recent collection, one arresting garment evokes a bird of paradise or exotic flower with its bursts of shocking pink and pale yellow. The fabric of the full skirt is layered with feather-like shapes, and the ribbed knit top sculpts the body like a second skin. Collins is not shy of colour or extravagance, yet there is always an element of classic simplicity present in her silhouettes.

Sabrina O'Cock

opposite
detail
Autumn/Winter 2006
knit cashmere, merino wool,
and cotton wrapped elastic

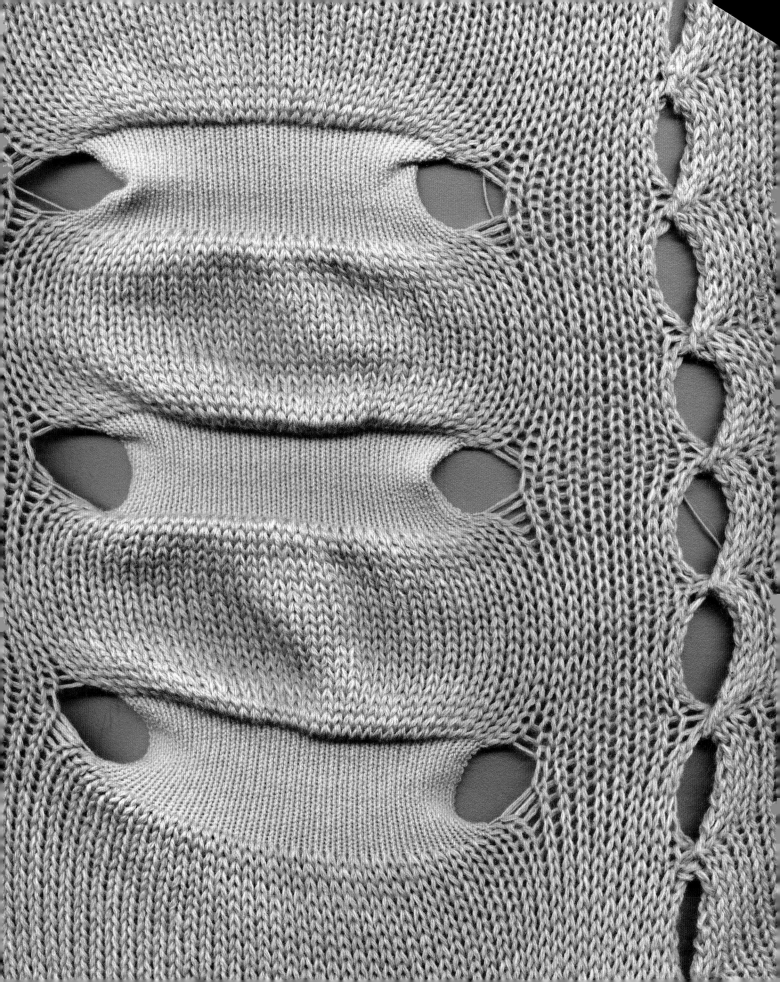

Spring/Summer 2001
knit cotton wrapped elastic
with silk chiffon top and
knit silk and cotton with silk
organza skirt
photo: Guy Caspary

opposite top left
Autumn/Winter 2001
knit merino wool and angora
with deerskin
photo: Dan Lecca

opposite top right
Autumn/Winter 2002
knit merino wool, cashmere,
angora, and rayon with
cotton gingham
photo: Rudy Martinez

opposite bottom
Autumn/Winter 2005
cotton shirting with knit
wool, rayon, and cotton
embellishments

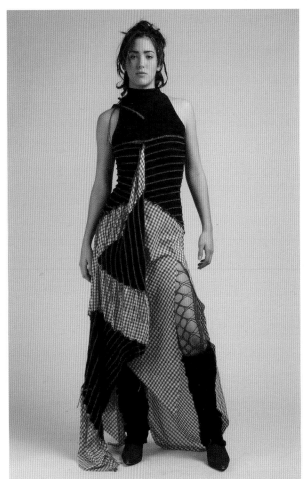

márcia**ganem**

opposite
Madame Pele
polyamide fibre
photo: Ricardo Fernandes

left and right
Ixchel
polyamide fibre
photo: Ricardo Fernandes **159**

Designer and craftswoman, Márcia Ganem is consistent in the creation of experimental, sculpturally constructed textiles, which stem unconditionally from an appreciation and understanding of native tradition. Tapping into the history and ambience of her hometown, Bahia, Brazil—a place defined by the convergence of different cultures and communities—she poetically translates personal perceptions and experiences into a body of work that is striking in both its aesthetic and context. Visually impressive and ingeniously crafted, her work goes beyond itself, adopting the essence, character and meaning of the environment in which it was conceived.

Remaining true to traditions permeating her surrounding community, Ganem has, for some time, collaborated with the lace-workers of Saubara, making the most of their exceptional abilities and age-old skills. Empathising with the history of their craft, she seeks to highlight notions of quality hand-craftsmanship bringing together the conventional and contemporary in collections that combine traditional methods of production with modern materials. Maintaining the form of their work but incorporating, for example, the fabrication of Landuti lace—a traditional, indigenous craft—with polyamide fibres, she is able to create luxurious, intricate fabrics that appeal to a far-reaching, intercontinental market.

The daughter of a seamstress, fashion and textile design comes naturally to Ganem, who, seeking influence from a disparate selection of sources, is dedicated to the creation of work rich in meaning and allusion. Infused with the tribal, mythological and musical, her work is often highly symbolic. Continually returning to themes of myth and motherhood, she pays homage to notions of femininity and fecundity, alluding to the concept of germination and new growth as a metaphor for her working processes. Her creations are constructed upon mannequins so that the fabrics, which are built up, can 'grow' to become a 'body'. Ganem is thus able to watch the cloth transform, shaping and sculpting it to suit. Working with a varied colour palette, ranging from muted shades of yellow, ochre, honey and gold to a more vibrant medley of rich blues and pure, crisp whites, Ganem produces textiles heavy in detail and surface texture. Delicately beaded materials are combined with sheer fabrics and coloured threads and fibres to produce an assortment of unique, inspired garments. Interwoven with a selection of semi-precious stones, these bejewelled creations are sculptural in their aesthetic, appearing like elaborate works of art for the body.

BC

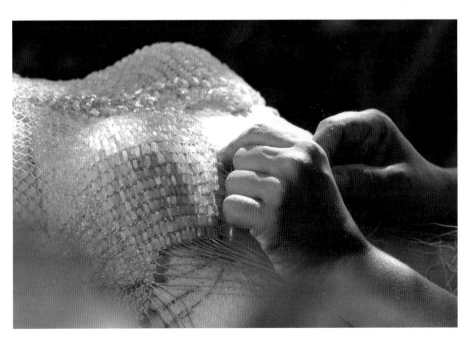

top
Juacema
polyamide fibre
and stones
photo: Ricardo Fernandes

bottom
work in progress
polyamide fibre
and stones
photo: Ricardo Fernandes

louise**goldin**

constructed

Having been awarded the coveted New Generation sponsorship by the British Fashion Council following her graduation from Central St Martins, Louise Goldin showed her Autumn/ Winter 2006 collection at London Fashion Week. Her fashion signature is, unusually, a completely knitted and constructed look—a combination of many different knit structures and textures, together with her distinctive hand-worked macramé-knotted accessories.

Her MA show featured a lean body conscious silhouette created by close fitting lightweight knitwear layers—sweaters, macramé leggings and longer sweater dresses—which provided a foil for the highly elaborate openwork knotted pieces worn over them. Not all the knits were plain, some featured uneven stripes made with slip stitch effects, mixing the subtle warm colour palette of greys and browns. Further visual interest and movement away from the body was created by the long fringes from the heavy cotton used in the macramé work, some of which was assymetrical in construction. This striking contrast between soft knitting and hard tight knotting was also incorporated into the same garment in some cases, utilising the macramé as inserts. Strong visual impact and technical prowess has led to two awards and the collection being sold in Selfridges and exhibited in Italy at the Pitti Filati Yarn Fair. The second collection, inspired by the work of Antonio Gaudi, has continued in the same

vein, adding more volume in curved shapes draping away from the body at shoulders and hips, and using yarns such as lurex, and viscose combined with cashmere in stitch textures, with a vivid blue accent.

Goldin has full control of the fabrication, textures and volumes though she collaborates with a knitwear manufacturer in Brazil to produce many of her fabrics and garments, a connection made during her previous experience designing knitwear for Brazilian label Tereza Santos. She has also designed for Rafael López, and had a work placement with Julien Macdonald, known for his innovative knits, whilst studying for her first degree. All of these influences come together in Goldin's mixing of weights, gauges, structures and shapes derived from experimentation on manual knitting machines. Her ability to work with sophisticated industrial knit production and traditional hand-craft techniques, which have rarely been used in fashion, signals the continuing development of a body of work which fuses both high and low technologies.

SB

opposite
Autumn/Winter 2005
cashmere and lurex knits
with cotton macramé
photo: Chris Moore

shirin**guild**

On seeing Shirin Guild's clothes for the first time, it is the combination of their volume, drape and texture which is striking—they have to be touched, felt next to the skin, above all worn. The embodiment of the clothes and their interaction with the human form is what marks out Guild's sophisticated fashions. These are clothes that flatter through movement of fabric, the line falling from the shoulders, which can be worn as well by the small or tall person, of any shape and size. Originally inspired by the voluminous traditional costumes of her native Iran, particularly for men, Guild has taken these and modernised the geometry of the oversized shapes and forms, but maintained the connections with Iran's ancient landscape, buildings and textiles through surface textures which subconsciously seem already lived in and comfortable to wear. The fabrics may be slightly faded, with deeply ingrained textures, or subtly coloured with lively movement but always highly appealing to our senses. The whole effect is one of pared-down simplicity, but with minute attention to detail.

Guild starts a collection with the textile development, which means sourcing both yarns and fabrics, as each season consists of half knitwear and half woven pieces. This is a crucial stage for Guild, where she visualises the final mood she wishes to create through colour and texture: subdued neutrals are sometimes offset with deep orange hues. She has a unique and close relationship with Italian fabric weavers Solbiati, the largest supplier of linen fabrics in the world, and communicates her vision directly with the firm's designer Carlo Lavazza, by discussing with him the qualities of her current inspirations, perhaps a kilim or old Iranian 'mountain cloth' fabric made of stiff goat's hair woven into very narrow strips. Through this liaison, new textiles emerge, such as her unusual damask patterned linens which have a much heavier drape than conventional plain weave linen. True to the ethos of the much-used textile, one fabric for Autumn/Winter 2006 exhibits, tongue-in-cheek, a woven-in pattern of stitched and mended cloth.

There are many unique fibre blends—only natural fibres are considered for the Shirin Guild label—including cashmere with linen, hemp, or steel and silk, and high twist wools which contrast and soften the hardness of linen and metal fibres. However, there is also a sensitivity to English traditions. For Autumn/Winter 2006, pure wool fabrics have been developed using two different twist yarns in a classic 'Prince of Wales' check, but then felted and milled to roughen the surface and create an undulating fabric effect. These fabric finishes are of equal importance to the actual textile construction in achieving the desired result.

left
Spring/Summer 2003
striped paper sweater and
striped indigo linen
fisherman pants
photo: Robin Guild

right
Spring/Summer 2004
steel and silk vest and abba
jacket worn with linen pants
photo: Robin Guild

165

shiringuild

The knitwear, which like all the garment production, is made in the UK, often features unusual fibres such as bamboo, and especially paper yarn, used in four collections to date, sourced from Japan. This is made from thin strips of paper which are twisted into heavy yarns, but when knitted and worn, the initial stiffness gradually softens into a silk-like touch. Linen is also extensively used for summer knits, less commonly found elsewhere. The well-known Shirin Guild wide rectangular shapes for knitwear have been recently developed into jackets which mix knits with woven fabrics, and are asymmetrical in form, pieced together in garment sections, with raw edges to the weaves.

In addition to her in-house design and production team, Guild commissions special pieces from individual designers, such as felted and dip-dyed knitted accessories by Caterina Radvan, hand-woven shawls by Tali Meshorer and hand-embroidered details by Karen Spurgin.

Through the subtlety of each season's design developments, Shirin Guild's silhouette and fabrications slowly evolve to always embrace the new, whilst respectfully drawing on the ancient past for inspiration and a sense of continuity.

SB

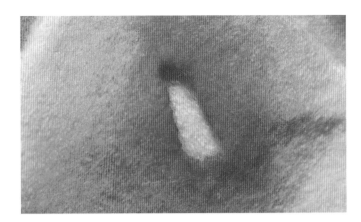

opposite
Autumn/Winter 2004
hand-felted and dip-dyed
wool/cashmere sweater and
leg warmers by Caterina
Radvan for Shirin Guild with
crushed wool pant panel
photo: Robin Guild

top left
detail of mitten by Caterina
Radvan for Shirin Guild
photo: Richard Pereira

top right
detail
Autumn/Winter 2005
wool and steel skirt
featuring woven design
inspired by African pots
photo: Richard Pereira

bottom
Spring/Summer 2004
printed silk by Jean Ensell
for Shirin Guild
photo: Richard Pereira

kanako**kajihara**

constructed

Using stretch yarns, linking devices and a host of different cutting and layering techniques, Kanako Kajihara creates a selection of single, double and multiple layered cloths, woven with both computerised looms and jacquard. These are often heavily detailed, drenched in a colourful array of flowing patterns and symbols. Inspired by both the organic forms of plants and animals, and man-made buildings and technology, her work is an effective union of synthetic and natural materials. Such juxtapositions allow for an inspired array of textiles that above all else remain true to the environment in which they were conceived.

Combining textile and fashion design, Kajihara has conceived a body of work which, inspired by natural methods of survival, allows for personal adaptation and modification. Creating a collection that enables the wearer to alter the pattern or colour of their clothing to complement their mood or environment, she prioritises flexibility and versatility. Organic-looking textiles made from materials such as Japanese paper thread and synthetic fibres form innovative garments that can be wrapped and layered in a variety of styles to suit.

Working with a colour palette that combines natural, earth-inspired tones with a brighter, more vibrant selection of golds, pinks, blues and greens, Kajihara openly references the opulence of the Guatemalan jungles, highlighting the beauty of the native flora and wildlife. The designer explains, "I spent some time travelling there and the bright colours and strange plants made such a strong impression—I wanted to feed these experiences back into my textile designs. Often people use colours to make a statement, using dark colours to conceal themselves and vivid colours to stand out, for example. This is something that is reflected in nature."

Using time spent working alongside the world renowned Issey Miyake to develop and hone her skills, Kajihara has been able to build an impressive textile portfolio that speaks of both progress and individuality. Her work, rather than following the fickle trends of fashion, is both enduring in its aesthetic and innovative in its technique.

BC

opposite left and right
Spring/Summer 2006
Royal College of Art
graduate collection
silk, cotton, polyester
and viscose

laineykeogh

constructed

Lainey Keogh has been realising garments both whimsical and sophisticated since 1987 by reinvigorating traditional Irish handwork and sculpting knit and crochet textiles into quintessentially modern forms. She was discovered by retailer Marianne Gunn O'Connor in the mid-1980s, and conceived her first collections by hand-spinning sheath dresses and scarves out of a home studio. Though the majority of Keogh's garments are still hand-made, the designer has employed a team of Irish knitters to produce all of her creations in Dublin since the founding of her label in 1994. At its most reductive, Keogh's portfolio is a fantastical revival of Irish lace traditions imported from Venetian and Spanish handwork epicentres in the early nineteenth century. Irish lace is typically crocheted and is distinguished by a square mesh ground dotted with stylised bas-relief roses or leaves. Variations are often named for the region from which they derive, and Keogh has utilised the two most popular mutations, Carrickmacross and Limerick, throughout her presentations.

Keogh vocally advocates 'friendly' fabrics in contemporary fashion, and is dedicated to the eradication of sweatshops and environmentally-unfriendly textile processes. With cashmere, wool, cotton, and synthetics lurex and polyamide amongst her favourite fibres, Keogh's politically conscious predisposition has prompted multiple variations of faux fur

opposite
Spring/Summer 2003
cords, feathers,
crochet lace
photo: Jackie Nickerson

Autumn/Winter 2003
cashmere with metallic
embroidery threads
photo: Jackie Nickerson

171

laineykeogh

trimmings and decorations, from a rust acrylic pile halter top for the Autumn/Winter 1998 London runway to a seasonal pageant of fibrous polyamide 'hair' tabards. Keogh's only notable use of animal raw materials was in her Amazonian collection of Autumn/Winter 1999, which offered a multitude of "Woolly Women" shrouded in cashmere and mohair knit gowns, beetlewing capes with lurex pile, and collapsible metallic knit chain mail.[1]

Ethnographic or spiritual overtones are not atypical of Lainey Keogh's conjurings. Her thematic interpretations frequently tribute the aesthetics of African or Asian cultures, from a macramé beige linen shell for Spring/Summer 1998, to blistered knit kimono and jellaba-style robes with embellishments that mimicked traditional court brocades on the Autumn/Winter 1998 runway.

The most potent Keogh concoctions are ethereal and tangible, resplendent yet simplistic. A review of her 20-year-old exploration of knitting and crochet intimates Keogh's fascination with the fibre as a raw material and elucidates her mastery at twisting and knotting these threads into dynamic three-dimensional forms.

EDC

jurgenlehl

Although born in Poland and of German nationality, Jurgen Lehl long ago found his spiritual home in Japan. Having first travelled there in the late 1960s to work for a textile company, he set up his ready-to-wear clothing business in 1972. He has fused Western and Eastern sensibilities in a rare movement against the fashion tide which has since flowed from Japan to the West. However, his clothes, although sold in over 40 outlets throughout Japan, are little-known outside, as there are no fashion shows and no publicity material to attract attention and only a handful of shops selling his clothes in Europe and America. He is happy for things to remain that way, preferring a local to global scale of operation.

Lehl is a consumate master of textile design, exhibiting a refined aestheticism more in tune with nature's cycles than with any fashion trend. His textiles reflect the earthiness of his inspiration, found largely in the natural landscape—collecting shells, stones and pebbles from the beach, many of which are used as fastenings in the clothes and accessories, and most recently have become a new range of jewellery. He has travelled extensively in India, Laos, Vietnam and China, as well as remote areas of Japan, taking photographs of the land (such as patterns in mud and sand, falling leaves and snow), and sourcing special techniques for his textiles.

Lehl is a champion of the hand-craft skills which create textiles and other artefacts in these regions (he also designs and makes ceramics and hand carved furniture) and prefers irregularity over uniformity.

Jurgen Lehl fabrics are increasingly complex—often developing several processes in one fabric, to achieve the effects he is seeking. A single length of cloth may have undergone each stage of production in a different country—perhaps woven in India, hand-stitched in China and *shibori* dyed in Japan, where it would also be manufactured into garments. The fabrics are all created from natural fibres, and dyed with vegetable-based natural dyes, which imparts a depth to the textured surfaces and deliciously soft, rich palettes of browns, purples and greens.

Love of the organic, non-repeating patterns of nature inspires both weaves and knits, which form about half of each of the seasonal collections. For example, knits will be designed with uneven fine multi-colour stripes, or irregular rib patterns, with colour changes woven subtly throughout the length of a coat or dress. Woven fabrics are pleated and dyed in a manner reminiscent of Fortuny, or made especially from extremely fine high twist yarns, which gives a characteristic slight unevenness to the surface, even in a plain fabric. Every fabric merits close attention, revealing fine

opposite
Spring/Summer 2006
reversible circular quilted
jacket in silk
photo: Sion Parkinson
courtesy Sandy Black

details, even in machine manufactured textiles which a lively surface and no obviously discernable repeats. Much of this comes from the combinations of yarns and fibres playing against each other, or from high twist 'singles' yarns used in knitwear to deliberately bias and twist the fabric.

Jurgen Lehl clothes have a timeless but contemporary appeal, utlising simple nonetheless elegant shapes, working in harmony with the fabrics, seen, for example, in the hand-quilted silk jacket, Winter 2005, with circular stitch patterning, reminiscent of a raked Japanese garden that is echoed in the cut of the jacket fronts. Textile designs are never repeated, but continually evolve, the company team learns and experiments with every new season, to offer their loyal customers something fresh each time. As Lehl believes: "Anything, anywhere can be a source of inspiration, but it does not come to fruition without the strong wish for new experience."[1]

SB

Spring/Summer 2006
hemp and cotton
appliqué dress
photo: Eva Takamine

opposite
Spring/Summer 2006
silk jacket with raw
edge seaming
photo: Eva Takamine

missoni

Missoni, hailed for a succinct amalgam of artistic and industrial genius in the form of lush knits and masterful manipulations of pattern and palette, has been a family-owned and run operation for over 50 years. Ottavio and Rosita Missoni's early manipulation of the Raschel Machine, used historically to produce open shawl patterns, provided their garments with uniquely airy knit formations nostalgic of folk motifs and ethnic tapestries that would later become their global namesake. The first collection that featured the inset "Missoni" label, entitled *Milano-Sympathy*, was sold at the Rinascente boutique in Milan in 1958. Early proponents of the Missoni style, which comprised fluid shirt dresses, tops, palazzo pants, and sweaters of varying knits and kaleidoscopic patterns, were Anna Piaggi, then-editor of *Arianna* magazine, who dubbed the look "haute boheme", and American fashion icon Diana Vreeland.[1]

The garments demonstrated a keen understanding of the post war transformation of high fashion from formal, structured French couture to a more casual jet-set wardrobe that, by the 1960s, was associated with Italian style. The popularity knitwear had enjoyed in the early twentieth century had waned since the Second World War, but Rosita Missoni's passion for the 1910s aesthetic of Gabrielle "Coco" Chanel, coupled with a revived demand for comfort and ease of care, have assured the couturiere's foundation garments of knit cardigan suits and shift dresses a secure place in fashion in the latter half of the century.

Although Missoni's interchangeable separates were derivative in cut of historical sportswear staples, their exquisite palette and use of new synthetics and natural fibre blends built their reputation as a leading fashion house by the late 1960s. In 1967, the Missonis debuted a collection at Florence's Pitti Palace in sheer knits worked through in varying stripes and juxtaposing shapes with colourful rayon and gold lame threads. These appeared on models that were nude beneath, prompting the outrage of the exhibition's organisers, but a simultaneous surge in sales for the duo's cutting edge creations.

Ottavio's distinctly complex patterns, reminiscent of the simplistic colour block forms of Russian Constructivism, fused brilliantly with Rosita's use of lurex and lame threads, recalled the pioneering attempts in metallic upholstery of the Bauhaus designers. The sheer, if explosively hued and gilded Missoni garments of the late 1960s provided a formidable launching point for the company, and as the only Missoni inclusions in Germano Celant's groundbreaking 1994 Guggenheim exhibition *Italian Metamorphosis, 1943–1968*, even serve as some of the house's most recognisable innovations.

opposite
Spring/Summer 2006
viscose blend double
jacquard sweater

Autumn/Winter 2002
zig-zag and space-dyed
mohair/wool blend
pieced knit
photo: Anthea Simms

constructed

As Missoni moved into the 1970s, knit configurations became more advanced and their patterns more frenetic, with favourites such as the *Patchwork* collection of 1971, which featured several different knits quilted together, as well as zig-zag, honeycomb and flame undulations that replaced bold stripes or diamonds. These signatures were made more recognisable by the Missonis' experimental space-dyeing technique, which dipped threads intermittently into dye baths, spotting them to affect a bleeding *ikat*-style colour composition.

This handwriting has been fortified and elaborated under the creative direction of the couple's children, Angela and Luca, since the late 1990s. Angela Missoni alternately texturises the Missoni womenswear template, through hyperbolic looped *bouclé* sweaters, or flattens it, impressing chromatic psychoactive prints on sheer chiffons and organzas. Concocting lavish suits of Missoni-tuned knit patterns and jersey dresses printed with the oscillating tempos engineered by her father, Missoni successfully re-articulates the timelessness of the brand's original aesthetic.

EDC

anne**maj**nafar

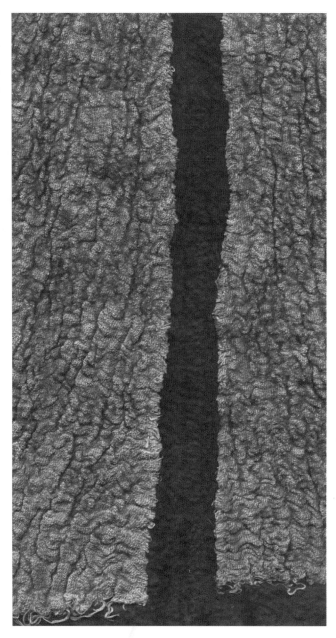

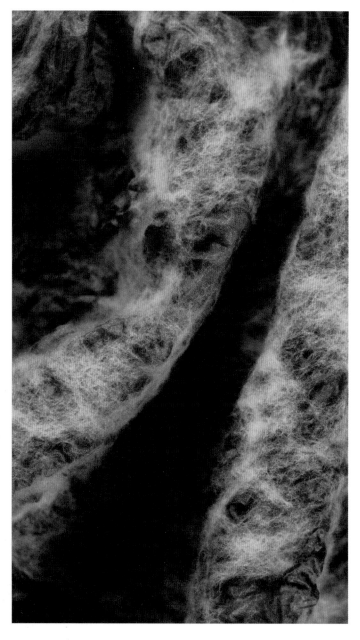

The Gods We Had Before
Autumn/Winter 2005

opposite left
detail
The Gods We Had Before
Autumn/Winter 2005
one piece shaped jacket
panels, merino wool felted
with base fabric klotzel

opposite right
detail
The Gods We Had Before
Autumn/Winter 2005
merino wool felted with base
fabric silk pongee

Part of a new generation of designers rebelling against mass-production and uniformity, Anne Maj Nafar has dedicated her work to the pursuit of a more holistic practice. Relinquishing technological textile production, Nafar seeks out methods of creation that are less predictable and embrace the irregular. Strongly influenced by retro-design and those who 'do-it-themselves', she has developed a body of work that speaks of natural evolution and quality hand-craftsmanship, upholding the theory that beauty can be found in imperfection. In contrast to the impersonal uniformity of modern technology, her work combines the raw and untouched with luxurious silk fabrics in a statement of the unification of "original basics and modern refinement".[1] Linking archaic images of the past to the fashion of today, Nafar integrates hand-felted fabrics with natural materials such as animal fur, organza, ponge and Klotzel linen and cotton.

Through a very hands-on working process dictated by chance, Nafar has fine-tuned the production of an entirely organic textile aesthetic. Using the nuno felting technique—consisting of wool fibres felted together with a woven material, such as cotton, silk or linen—as well as Merino and knitted wool, she is able to create elemental looking textiles enriched with touches of colour and elaborately tactile surfaces. Envisaged in

a natural colour palette of earthy browns and creams, the felts are enlivened through the incorporation of textured fibres woven freely into the base materials. Transcending the innate human impulse to control, Nafar enables her fabrics to 'grow' and develop with a will of their own, encouraging natural 'imperfections' and patterns. The apparently seamless garments are a result of a real, physical process—a process that can only partially be controlled. "Body and mind, the felt-maker and the product, become like one, and the beauty lies not only in the final product but in the process itself."

Inspired by the German artist Joseph Beuys and the healing ability he ascribes to felt, Nafar has been persistent in her quest for a more spiritually driven form of textile design. Striving towards beauty and meaning through the formation of organic silhouettes, basic, simple lines and elegant, flowing shapes, Nafar is consistent both in her understanding of aesthetics and in the production of exceptional textiles.

BC

opposite top
Anne Maj Nafar performing
the felting process,
February 2006

opposite bottom
detail
The Gods We Had Before
Autumn/Winter 2005
superfine merino wool 15%
camel hair, 15% alpaca,
10% mulberry silk felted
with base fabric silk chiffon

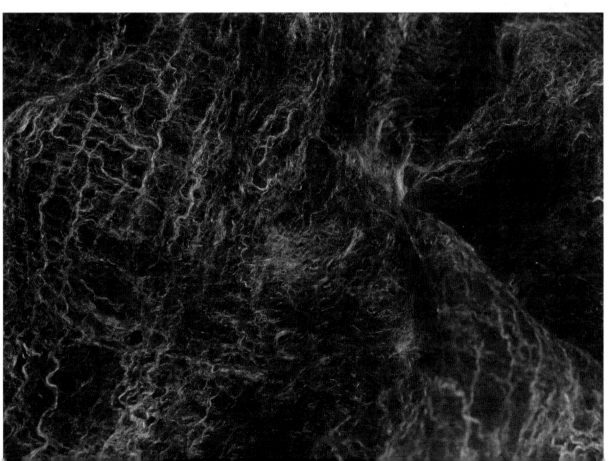

left
The Gods We Had Before
Autumn/Winter 2005

right
detail
The Gods We Had Before
Autumn/Winter 2005
merino wool loosely knitted
and felted with base fabric
silk organza

left
The Gods We Had Before
Autumn/Winter 2005

right
detail
The Gods We Had Before
Autumn/Winter 2005
felted merino wool,
showing integral pocket

185

annemajnafar

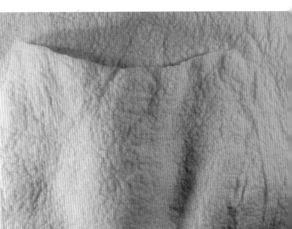

hikaru**noguchi**

constructed

Constantly searching the world for the best materials, Hikaru Noguchi isn't afraid to take risks, experimenting with time-consuming, unusual techniques in knitwear that larger manufacturers are hesitant to try. Her success can be credited to her eclectic mix of traditional knits, embroidery, and Fair Isle patterns, that come together to create a distinctive style that is to be found in fashionable wardrobes the world over. The use of the highest quality yarns and materials, which she sources in Scotland, Italy, and elsewhere, is a distinguishing characteristic of each piece in her collections. Whether using lurex, a type of metal yarn, to create the three-dimensional zig-zag effect in her "Peak" scarves, or mixing antique kimono fabric with her own knitted fabric, innovation is a hallmark of her design ethos. The first to come up with the method of knitting loosely spun slub yarn into a base fabric to create her best-selling shaggy scarves, this technique has been duplicated by so many other designers that it is now regarded as standard.

Finding her inspiration in flea markets, nature, people-watching and art, Noguchi tries to convey a myriad of textures and shapes with each collection. Her love of tweed, described by Noguchi as a "well-worn sheep farmer's jacket", is prominent in her designs.[1] Using the best Italian tweed yarn, a mixture of wool, alpaca, and silk, the result is a hybrid of softness and light with a country feel. Lambswool is another favourite material due to its versatility, the look and feel of the finished product depending on if and how it is washed.

Always searching for fresh ways to manipulate texture, Noguchi's latest interest lies with utilising traditional knitting techniques with yarns that are not typically used in that manner. In 2006 she launched a new range called "Le Mouflon Toque", including garments using fabrics which are predominantly sourced in South Africa. One of those fabrics is traditional discharge-printed cotton that can be traced to mid-nineteenth century European settlers in Africa who wore printed textiles with strong Asian influences. Whether designing intricately patterned fabrics or spinning new textures out of ordinary yarn, Noguchi continues to keep knitwear looking contemporary.

JT

opposite
Autumn/Winter 2005
cut and felted wool scarves,
loosely woven wool

Autumn/Winter 2005
knitted wool with jacquard
patterning
photo: Sion Parkinson

opposite top right
Autumn/Winter 2005
knitted relief patterning in
wool and lurex
photo: Sion Parkinson

opposite bottom left
Autumn/Winter 2005
inlaid space-dyed wool slub
on knit wool base
photo: Sion Parkinson

opposite bottom right
Autumn/Winter 2005
woven patterned wool fabric
photo: Sion Parkinson

jessica**ogden**

constructed

A proponent of 'salvage' fashion, Jessica Ogden has explored the rejuvenation of historic garments, textile treatments, and craft traditions from Japan to New England. Although her own collection was launched in 1993, Ogden's folksy ensembles have only gained global recognition in the past five years. Her late 1990s creations featured recycled, bleached and top-stitched denims, as well as photocopied cotton canvases that accentuated both a tactile astuteness and a passion for surface design.

Ogden's Autumn/Winter 2001 collection, examined deconstruction as a means of recoding the vintage garment, with the Rhode Island School of Design graduate showing tattered brocades and shredded silks, disfigured with cheese graters, hammers and nail guns. Even at the height of the vogue for both distressed textiles and vintage clothing consumption in global fashion, Ogden's portfolio had a distinct character. Using friends and relatives of all different artisanal occupations as the mannequins for that 2001 collection, the London-based designer conveyed her vision effectively as one celebratory of a versatile community of female artists. This collection, as with all those to date, championed a reinvigoration of craft, and more specifically of women's traditional artisanship.

Ogden's creations, through the use of both brightly coloured textiles and natural fabrics intended for comfort, exude a child-like innocence, but are also motivated by a feminist dialogue implicit in the revival of various women's textile traditions. A strong Japanese influence marked Ogden's Spring/Summer 2002 presentation, and a two-piece kimono-wrapped ensemble, printed with broad, abstract strokes was one of her most telling creations; the roller print flaunted a distinct likeness to Japanese Sumi-E calligraphy, which was assimilated from China and integrated into the studies of young Japanese girls in preparation for marriage and womanhood.[1] The same collection introduced the quilted garment into Jessica Ogden's repertoire with worn crazy-pieced tartan and pastel plaid sundresses. These early quilted designs mark the beginning of Ogden's persisting attention to the historical practices of American quilting, which in later collections was further refined through more expert seaming, refined palettes and fluid patchwork.

Ogden presented an informed slice of Americana for Spring/Summer 2005, from white-work capelets (white embroidery or appliqué on a white ground) and 'whole cloth' asymmetric wraps to a 'block' quilt that revived the popular nineteenth century Turkey Red hue. One of her garments even reflected the Indian Kutch checkerboard pattern that

Kieta Stripes
Spring/Summer 2005
antique checkerboard
quilted cotton and gingham
pieced with printed cotton

migrated to Colonial quilting. Ogden's *Madras* resort line (co-designed by APC founder Jean Touitou) launched in July of 2004, and has since infused her collections with a range of prints and quilting techniques traditional to India.

In recent seasons, Ogden has honed her aesthetic and continued to evoke a new dialogue for craft revival. Autumn/Winter 2005's use of Colonial quilt patterns like "Harlequin", "Diamond in the Square", or combination "Sunburst" and "Pinwheel" repeat all demonstrate Ogden's continued allegiance to the traditions that governed American quilting. One of the most atypical and innovative designs of this presentation is an elongated tunic with drawstring waist and stylised leaf and bird appliqués. The rustic design potently homages mid-twentieth century Finnish and Scandinavian designers like Armi Ratia of Marimekko, who, like Ogden, have reconstrued traditional textile arts (in Ratia's case block, roller, and screen printing) in order to offer a contemporary take on femininity and nostalgia.

EDC

claretough

Sex and knitting, two terms that might be considered incongruous, are frequently brought together in discussing Clare Tough, a young British knitwear designer who quickly caught the attention of the fashion press after her 2004 MA graduation show. Hailed by British *Vogue* as "the future of British fashion", Tough's designs re-interpret traditional knitwear shapes, sectioning the wearer's body into fragments of contrasting colours, slashes and cut-out sections.

Tough's textiles combine hand-knitting, machine knitting and crochet, in mixes of linen, cashmere and cotton, in a patchwork of gauges from tight to free and looping. Textures and colours are juxtaposed; rough brushes up against smooth, chunky against delicate, and muted against lurid. However, every contrast is also a deftly achieved synthesis. Tough's creative process begins with the draping of fabrics across the body, which she then photographs and, using a process of photo-montage to experiment with arrangements of shape and colour, produces whorling sections of fabric and cut-out. The combination of solid materials and revealing holes suggest the craft of knitting itself—a balancing act struck between space and non-space. As shoulders, breasts and hips poke through gaps left by the designer, in a play of revealing and concealing, the knitwear's erotic potential becomes apparent.

Indeed, it is a tradition of erotic garments that Tough draws upon. For her graduate collection the designer took her inspiration from hosiery and fishnet stockings. Stretching fabrics around bodies, and copying the patterns that this movement created, she then attempted to develop fabrics that "emulated these patterns without actually being stretched."[1] The appeal of this aesthetic is accentuated by the designer's collage of colours, as juicy and tempting as a fruit basket.

Similarly, her Autumn/Winter 2006 collection included pieces based on the traditional corset. Using a dark, glamorous palette and heavy, glossy yarns, Tough fashioned tight-fitting garments with outlines mapping the contours of the breasts in bright gold lurex, before coursing down the centre of the body in vertical strips, creating a close silhouette. Tough's creative manipulation of her craft is in evidence here, too; she has developed a technique of knitting yarn into gold chains. Crochet, she explains, is by nature simply a series of chains. It is these small acts of simple alchemy which further mark her out as a designer who has turned the dowdiest of fabrics into the most luxurious garments.

LMF

opposite and overleaf
Autumn/Winter 2005
hand-knitted or crocheted
wool, cashmere, cotton, silk,
linen, acrylic and mohair

biographies

sandy**black**

Editor and co-author Sandy Black is the author of *Knitwear in Fashion*, and has contributed to numerous other publications including *Textiles for Protection*, and the magazines *International Textiles* and *Selvedge*. She curated the exhibition *The New Knitting*, which toured the UK in 1998 and 2000–2002, and before becoming a lecturer, ran her own internationally-selling knitwear label. She is currently a Reader in Fashion Design and Technology at the London College of Fashion, University of the Arts.

elyssa**da**cruz

Co-author Elyssa Da Cruz is a fashion historian and journalist residing in New York. She is currently a Research Associate at the Costume Institute, where she served as co-curator for 2004's *WILD: Fashion Untamed* and co-authored the accompanying exhibition catalogue. Da Cruz is a part-time professor at The Cooper-Hewitt Graduate Program in Decorative Arts under Parsons/The New School, and has contributed to *The Encyclopedia of Clothing and Fashion*, as well as publications such as *Zink*, *CITY*, and *Elle* Canada.

stephanie**aman**

Born 1979, Chichester, UK, lives and works in London.

With an innovative fusion of sharp, edgy cuts and classic embroidered textiles, British newcomer Stephanie Aman piqued the interest of designers and critics alike while still at college—receiving the coveted Levi's Bursary Design Award in 2003 and soon after attracting attention for her 2005 Royal College of Art MA show. It didn't take long for some of the biggest names in fashion to tip their hat to her, including Rei Kawakubo, whose prestigious London concept store Dover Street Market, immediately bought her entire graduating collection. While undertaking her BA Honours at Middlesex University from 2000–2003, Aman worked as Managing Director for established designer Cathryn Avison, designing for the fashion and embroidery collections each season. In 2006, Stephanie Aman set up her own fashion label, Aman Copland, with business partner Clare Copland, with whom she has released her first capsule collection as well as a limited edition lingerie collection.

manish**arora**

Born 1972, Bombay, lives and works in New Delhi.

Having been presented with numerous awards in his native India—including Most Original Collection at the 1995 Young Asian Designer's Competition in Jakarta—Manish Arora is currently one of the most celebrated designers of the Subcontinent. He is gaining acclaim in the international fashion scene as well, with the accolade for Best Collection at Miami Fashion Week bestowed upon him in May of 2005. Aside from being stocked at the Galleries Lafayette and Saks Fifth Avenue, he retails in 23 exclusive outlets in eight countries worldwide. The Manish Arora label includes an opulent line of couture specially sponsored by Swarovski, and a trendy line of trainers that goes by the name of Fish Fry—a collaboration between himself and sportswear mammoth Reebok. Fabrics and features included in the Arora *oeuvre* include silk appliqué, hand-embroidered sequins, brocade, lurex frills, raw silk, faux fur, chiffon, and tassels, combining European dressmaking traditions with Oriental textiles and techniques.

ashish

Ashish Gupta, born 1973, Delhi, lives and works in London and Delhi.

Following in the footsteps of countless fashion luminaries, Ashish Gupta graduated from Central St Martins to instant acclaim—securing prestigious Browns Focus of London as his initial stockist, and receiving the British Fashion Council's New Generation designer award three times. Gupta's original, high-quality garments seamlessly merge East and West, mirroring his own harmonious existence between London and Delhi—the cities where he respectively designs and has his clothing made. Gupta left Delhi after completing a Fine Art degree, arriving in London at 21. Initially studying Fashion Design at Middlesex University, Gupta later undertook an MA in Fashion at Central St Martins, graduating in 2000. Gupta's innovative designs for his self-titled label Ashish draw inspiration from many cultures, and employ intricate hand-crafting techniques such as Indian embroidery and beading, as well as unique fabrics designed by Gupta himself. All Ashish garments are manufactured and hand-finished in India where Gupta owns and operates a workshop.

ioannis**dimitrousis**

Born 1980, Thessaloniki, Greece, lives and works in London.

Following on from his fashion studies in his native Greece, Ioannis Dimitrousis studied for his degree in Menswear Fashion Design and Technology at London College of Fashion. His heavily textile-based work comprises a colourful combination of various knitted pieces, embroidery and appliqué, made into chunky scarves and accessories to complement his sports-inspired look. The designer relies on the use of hand-crafted techniques to combine new and used fabrics and yarns into unqiue, one-off pieces. Dimitrousis has an impressive array of work experience under his belt, including as a pattern cutter and machinist to a variety of high profile designers such as Roland Mouret, Jonathan Saunders and Louis de Gama. Now finding his feet in the London fashion scene, he has since launched his own bespoke mens- and womenswear fashion label in London. Having recently shown at London Fashion Week as part of the On/Off schedule, the designer is taking commissions for work both in London and Greece, where he has established a strong following.

john**galliano**

Born 1960, Gibraltar, lives and works in Paris.

Considered one of Britain's fashion icons, John Galliano moved to south London at the age of six. A 1983 graduate of Central St Martins, his graduation collection, inspired by the French Revolution and called *Les Incroyables*, was bought by Joan Burnstein of Browns, launching his career as a major player in the fashion community. Receiving widespread critical acclaim for his Latin-influenced, romantic and feminine bias-cut dresses and fusion of Mediterranean colour in textile design, he inspired loyalty among fashion insiders. He moved to Paris in 1992 and went on to become the first British designer to head a French couture house when he was made Chief Designer at Givenchy in 1995. A year later he was given creative control of Christian Dior. One of the most influential designers of our time, he is a four time winner of British Designer of the Year, in 1987, 1994, 1995 and 1997.

yoshiki**hishinuma**

Born 1958, Sendai City, Japan, lives and works in Tokyo.

Yoshiki Hishinuma gained his expertise in fabric treatments from his studies at the Bunka College of Fashion and his experience working for the Miyake Design Studio. In 1983 he received the Mainichi Fashion Grand Prix for young talent and in 1992 he introduced his own fashion label and has been showing his collections in Tokyo and Paris ever since. Hishinuma is especially known for his innovations in textile design, combining new technologies with traditional Japanese techniques. He received the Mainichi Fashion Grand Prix for a second time in 1996 for his applications of new technologies in combination with traditional Japanese *shibori* designs. More recent innovations include his signature laser-cutting technique, which was developed for his David Lynch inspired Autumn/Winter 2003–2004 collection. Besides this, he has designed costumes for theatre, film, dance and opera. In 1999 Hishinuma had a retrospective of his work at Gemeentemuseum den Haag.

abu**jani**&sandeep**khosla**

Abu Jani, born 1960, Mumbai, and Sandeep Khosla, born 1963, Amritsar, both live and work in Mumbai.

Abu Jani and Sandeep Khosla began their respective careers working separately for Xerxes Bhatena, designing costumes for the Bollywood film industry. Jani's costume design credits also include musicals such as Vivek Prabhakar's *Grease Lightning* and Alyque Padamsee's *Evita*. Khosla and Jani, having met through their mutual employer, formed their own label Abu Sandeep in 1996. Together, they attracted attention with their very first collection for the boutique Mata Hari, which was featured on the cover of Bombay magazine. Celebrity clientele include Amrita Singh, Jaya Bachchan, Dame Judi Dench and Dimple Kapadia, and their ensembles were featured in the 1999 James Bond film, *The World is Not Enough*. Considered India's highest honour, they were presented with the National Award in 2003 for the film *Devdas*, designing costumes for the main characters. Their recent foray into the world of ready-to-wear at the 2005 Lakme Indian Fashion Week offers their brand of luxury at a more accessible price point. Their lines are sold in boutiques from Bangalore to London.

kenzo

Takada Kenzo, born 1940, Himeiji City, Japan, and Antonio Marras, born 1961, Alghero, Sardinia, lives and works in Paris.

Kenzo left Kobe University and the literature studies prescribed by his parents to become one of the first male students at Tokyo's Bunka Fashion College, Bunkafukuso Gakuin. In 1964 Kenzo found work as a designer for Pisanti in Paris, before opening his first boutique Jungle Jap in 1970. Kenzo's early work borrowed heavily from traditional Japanese styles but it was his 'big silhouette' designs that drew worldwide attention. The Kenzo label became known for its use of striking floral and jungle prints, as well as a fusion of styles from across the globe—Kenzo Takada is often described as fashion's most prominent traveller. Kenzo is now owned by the luxury fashion empire LVMH, and has a children's line, as well as several perfumes and beauty products. In September 1999, he announced that he was handing over the reins of his fashion house to his second-in-command, Giles Rosier, who led the label until 2003, when Antonio Marras took charge.

Antonio Marras grew up surrounded by rolls of fabric in his grandfather's shop, and felt such a natural aptitude for fashion that he quickly moved into the industry without formal study. His passion for textiles managed to convince an entrepreneur from Rome to back him to create his first ready-to-wear collection in 1988, named *Piano Piano Dolce Carlotta*—after the Robert Aldrich horror movie of the 1960s. He made his haute couture debut under his own name in Paris in 1996, followed by women's ready-to-wear in 1999, and men's ready-to-wear in 2002. Under Marras' control, The House of Kenzo's collections have been in keeping with the fusion of cultures that Kenzo is known for, while adding his own touches of Sardinian folklore influence.

christianlacroix

Born 1951, Arles, France, lives and works in Paris.

Christian Lacroix studied Art History at the University of Montpellier, and then continued his studies at the Sorbonne and the École du Louvre in Paris, with the ambition of becoming a museum curator. He developed his interest in fashion in the late 1970s, and due to his friendship with Jean-Jacques Picart, a press attaché and advisor to several designers and haute couture houses, he landed a job at Hermès in 1978 and then with Guy Paulin in 1980. This invaluable experience was followed by a position at the venerable fashion house of Parisian designer Jean Patou in 1981. There he learned the skills of haute couture, eventually leaving to start his own house with Picart. Lacroix's designs took the fashion world by storm and he gained recognition for his exquisite embroidery and patchwork. Throughout the 1980s his inventive, flamboyant and technically astute creations gained him worldwide prestige. His eclectic influences, ranging from ethnic costumes to historical dress and popular culture, make his collections theatrical, memorable and highly influential. The recipient of numerous awards and honours, Christian Lacroix was the first designer to be admitted to the Chambre Syndicale for more than 25 years.

projectalabama

Natalie Chanin, born 1961, Florence, Alabama, where she continues to live and work.

Natalie Chanin founded Project Alabama in 2000 after completing a degree in Environmental Design from North Carolina State University, and some work experience in the junior sportswear industry and as a stylist and costume designer. Chanin's conceptual design background is balanced by the extensive marketing and finance experience of her New York business partner, Enrico Marone-Cinzano, which has guided the company to its unique position in the fashion industry. Now in its sixth year, Project Alabama offers a full womenswear collection as well as a line of men's and women's tee-shirts and home accessories. The label employs local artisans, called stitchers, to produce the lines, and in this way helps to foster the region's re-blossoming textiles industry. Bauhaus-based ethics and philosophy surround Project Alabama's work. Each collection consists of custom garments with a modern twist, emphasising quality of cut, detail and style. Each product is a labour of love, signed by its primary stitcher on a hand-made label—proof of the pride woven into the process.

threeasfour

Adi, born 1974, Tel Aviv, Israel, Angela, born 1972, Dushambe, Tajikistan, and Gabi, born 1965, Beirut, Lebanon, all live and work in New York.

The international design collective ThreeAsFour, emerging from different cultures and backgrounds, found a home in New York City's Chinatown. Adi and Ange, respectively of Israeli and Russian origin, and Gabi, originally of Palestinian descent, have been collaborating since the late 1990s, broadening the boundaries of fashion, art and entertainment. The designers, primarily known for their multi-media fashion shows and curvilinear design interpretation, have established and defined an individual style. ThreeAsFour is presently collaborating with Kate Spade on a collection of clothing as accessories, combining evolutionary ideas with their own vision of the classic 'American Dream'. The team launched As FOURDENIM in 2005 at the ThreeAsFour store in lower Manhattan, and a ThreeAsFour perfume for Colette.

driesvannoten

Born 1958, Antwerp, where he continues to live and work.

Maverick Belgian designer Dries Van Noten's idiosyncratic fashion foresight has earned him continual renown, with his memorable entrance onto the London fashion scene in the mid-1980s marking the beginning of his career as a vanguard couturier. Placing Belgium securely on the fashion map, Van Noten, and famed counterparts of the so-called 'Antwerp Six', hosted a group show in 1986 to immediate acclaim, with Van Noten walking away from the experience with his first three stockists—the eminent Barneys of New York, Whistles of London and Pauw of Amsterdam. Born in Antwerp in 1958 to a family of tailors going back three generations, it seemed natural that Van Noten, whose earliest memories involved design, would pursue a career in fashion—graduating from the esteemed Fashion Department of The Antwerp Royal Academy of Fine Arts in 1981. Subsequent to his successful London showing in 1986, Van Noten opened a boutique in Antwerp in 1989, Het Modepaleis, which today continues to stock his own men's and women's collections, as well as that of select fashion and accessory designers. In 1993 Dries Van Noten took an international turn, opening showrooms in Paris and Milan; in1997, he expanded his intimate empire further to exclusive outlets in Tokyo and Hong Kong. Notwithstanding a staunch refusal to advertise, Van Noten is currently stocked in well over 500 elite stores worldwide.

martinevan'thul

Lives and works in Amsterdam.

Martine Van't Hul's career was launched when the music artist Björk purchased several garments from the designer's MA graduate collection, at the Fashion Institute, Arnhem, for her world tour and for several publicity shoots. With a flair for detailed and complex fabric embellishment, Van't Hul moved to Milan, where she became an assistant to designer Ronald van der Kemp. As she moved into freelance design in Milan, she began designing fabrics for Larus Miani, whilst also designing her own collections between 2000–2003. Her collections were notable for their contrasting uses of synthetic and hand-made fabrics, and their 'unfinished' appearance, which the designer named *Mi-Confectioné*. After several collaborative projects, Van't Hul moved towards an art/fashion crossover, taking what she describes as a more "painterly" approach to fabrics in collaborating with artist Fleur van Maarschalkerwaart, a visual artist known for her digital drawing techniques, and embellishing her drawings with embroideries at the *Dutch Touch* exhibition in Paris and New York in 2005.

husseinchalayan

Born 1970, Nicosia, Cyprus, lives and works in London.

Hussein Chalayan graduated from Central St Martins in 1993 and launched his own label the following year. In 1995 he was awarded the Absolut Design Award and was named British Designer of the Year in both 1999 and 2000. Chalayan's conceptual practice uses film, installations and sculptural forms to explore perception and the realities of modern life. His work is presented in shows and galleries, whilst his clothing is available in boutiques in the world's major cities. He lectures regularly and has exhibited in museums worldwide, including the Victoria and Albert and The Hayward Gallery in London, The Kyoto Costume Institute in Japan and the Musée de la Mode et du Textile, Palais du Louvre, in Paris. Chalayan created costumes for Michael Clark's 1998 production of *current/SEE* and has designed for TSE Cashmere, Top Shop, and Marks and Spencer. In 2001 he was appointed creative director at Asprey and began to show his own womenswear label in Paris in 2002. The same year he launched a menswear collection and edited *No C* magazine in Belgium. A ten year retrospective of his work was exhibited at the Groninger Museum in The Netherlands in 2005. His recent forays into filmmaking landed him a spot as Turkey's representative in the 2005 Venice Biennale, where he showed his film *The Absent Presence*.

sangheechun

Born, 1972, DaeJeon, South Korea, lives and works in San Giovanni, Italy.

In 2002, Sanghee Chun came to the UK to study Fashion Design and Technology at London College of Fashion. Having already completed her BA in Clothing and Textile Design at Chungnam University, and spent five years working for high street retailers in South Korea, the designer was well equipped to complete her MA in London, graduating with a distinction. During her studies in London, Chun was production assistant for both Alexander McQueen and Tata-Naka (a young London-based fashion company), working on pattern-cutting, embroidery, sewing and dyeing. Her MA graduate collection showcased her blending of print with a range of cutting and construction techniques, earning her a position with Alberta Ferretti in Italy.

commedesgarçons
Rei Kawakubo, born 1942, Tokyo, lives and works in Tokyo

After graduating in Fine Arts from Keio University, Tokyo, in 1964, Rei Kawakubo worked for two years in the advertising department of Asahi Kasei, a manufacturer of acrylic textiles. Drawing upon this experience, she went on to found Comme des Garçons in 1969. The firm was incorporated in 1973, and launched menswear, knitwear, ready-to wear and furniture lines in the following decade. The first Comme des Garçons boutiques were opened in Paris in the 1980s, and the flagship Tokyo store followed in 1989. Given Kawakubo's unusual point of entry to the industry, with no formal training in fashion design, she conveys her design ideas to her patternmakers in words—which is perhaps telling, given the highly conceptual nature of her work. The designer's work employs a range of textile technologies, from traditional Japanese techniques, to the use of rubber and modern synthetic blends, as well as deconstructing fabrics to produce puckers, tucks, and gathers. Experimenting with form, her clothing is variously loose and layered, sculpted, and taut to the figure, creating unique and radical silhouettes which, because of their austere and deconstructionist aesthetic, have been described as 'anti-fashion'. Acclaimed by the media and the fashion industry alike, she was awarded the Mainichi Newspaper Fashion Award in 1983 and 1988, and the Parisian Ordre des Arts et des Lettres in 1993, among a host of other accolades; her work has been showcased in both solo and group exhibitions in Paris, New York, Tokyo, and other locations around the world.

carlafernández
Born 1973, Saltillo, Coahuila, México, lives and works in México.

Carla Fernández commenced her fashion career after studying Art History, specialising in Textiles, at the University Iberoamericana in México City, and continuing her education at the Sorbonne in Paris. On returning to México, Fernández refined her education in haute couture at the design school, Iberoamericana de Diseño. In 2003, she created her innovative design studio, Flora 2, which was developed to improve the lives of local artisans by combining their traditional creative knowledge with cutting edge fashion design. Fernández is heavily involved with a number of government projects in order to fully immerse herself with local communities, such as PROADA (Programme to Assist the Development of Craftsmanship), and works extensively throughout villages in México to grasp a greater understanding of local traditional dress and textile techniques. She currently works alongside Xavier Rodriguez, and also pursues work as an art historian and costume designer. Her work has received acclaim from journals such as *i-D* magazine, and has been shown at México City Week in London, and the Museum of Contemporary Art in Japan.

junkokoshino
Born 1939, Osaka, Japan, lives and works in Tokyo.

Having graduated from a degree in Fashion Design at Bunka Fashion College, Tokyo, in 1966, Junko Koshino went on to open her first boutique in Tokyo in 1966; stores in China, Paris, New York, and Singapore followed over the course of the next 30 years. Technical innovation is employed to seamlessly merge Modernist and Japanese traditions in her work, which combines extraordinary sculptural form with a minimalist palette. Aside from ready-to-wear and couture lines, both established in 1978, Koshino also runs lines of home furnishings and wigs. Across this diverse range of practice, and throughout her career, the designer has brought to bear a range of advances in textile manipulation, exploring her structural aesthetic through natural and synthetic materials alike. She has designed uniforms for sports teams and corporations, and costumes for a number of internationally renowned opera houses. Her impact, innovation and influence have been celebrated in exhibitions at the Museum of Chinese History in Beijing, in 1992, and at the Carrousel du Louvre, Paris, in 1994.

ezinmambonu
Born 1970, London, where she continues to live and work.

Ezinma Mbonu's varied career to date has been as vibrant and diverse as her own creations. After graduating from Central St Martins in 1993 with a BA Honours in Fashion Womenswear, Mbonu refined her distinctive style whilst undertaking an MA in Fashion at the Royal College of Art in 2003. During this period she also turned her hand to fashion styling for music videos, commercials and numerous editorials, appearing in prestigious magazines such as *The New York Times Magazine* and *Viewpoint*. Since graduating, Mbonu has designed costumes for Ridley Scott's *Kingdom of Heaven*, and worked as an assistant designer at Lezley George. Mbonu has also exhibited at the Oriki exhibition and gained distinguished media attention after appearing in *Fashion House*, a Channel Four production, in 2004, and making an impression on Giovanni Valentino himself. Mbonu currently showcases her own label at Cockpit Arts, a creative hub for British designers in London.

rowanmersh
Born 1983, Bournemouth, UK, lives and works in London.

Rowan Mersh, a 2005 MA graduate of the Royal College of Art, has already been widely recognised for his extraordinary, imaginative abstract forms and innovations. In 2004 he received the Haberdasher's Award, and his work *The Shape of Music*, a mixed-media sculpture, impressed the judges with its experimental construction and won top prize in the 2005 Nationwide Mercury Prize Competition. Since receiving his BA Honours in Multi-Media Textiles from Loughborough College of Art and Design in 2003, Mersh has worked in London as a designer and consultant for Karen Nichol, Kjaer Global, and Pieke Bergman, and meanwhile exhibited in *Raw Talent* at The Atlantis Gallery and *New Designers* at the Business Design Centre, Islington. His technical innovations in textiles lead to unique applications in mens- and womenswear, and his taste for sculptural work applies with equal ease to the runway and interior design.

issey**miyake**

Issey Miyake, born 1939, Hiroshima, Japan, and Naoki Takizawa, born 1960, Tokyo, where he continues to live and work.

Miyake studied Graphic Design at the Tama Art University in Tokyo, graduating in 1964. After a stint studying at the Chambre Syndicale de la Couture Parisienne in 1965, Miyake worked for Guy Laroche and Hubert de Givenchy in Paris, and then Geoffrey Beene in New York. Returning to Tokyo in 1970, he founded the Miyake Design Studio. Miyake's seminal technology-driven creations stem from one concept: that of creating clothing from a single piece of cloth. In the late 1980s, he began to experiment with new methods of pleating that would allow both flexibility of movement for the wearer, and ease of care and production. This eventually resulted in a new technique called 'garment pleating' for his 1993 Pleats Please debut collection. In 1994 and 1999, Miyake turned over the design of the men's and women's collections respectively to his associate, Naoki Takizawa. He continues to design *APOC* (A Piece of Cloth), a line that was introduced in 1999 in which the components for a fully finished woven or knit garment are made in a single process. Miyake has numerous stores in Tokyo, Paris, New York, and London. His New York flagship store, Tribeca Issey Miyake, was designed by Frank Gehry, and is the first to carry all seven of his collections, including the Issey Miyake Womens, Mens, Fete, *Pleats Please*, *APOC*, Me and HaaT lines. In 2000, after 30 years at the forefront of design, Issey Miyake handed over the reins of the Miyake Design Studio to his right hand man Naoki Takizawa, who has been with him since 1989. Takizawa studied in Japan at the Kuwasawa Design School.

yohji**yamamoto**

Born 1943, Tokyo, Japan, lives and works in Japan.

Famed for juxtaposing the simple appearance of his designs against the sophistication of their construction, Yohji Yamamoto never fails to attract both admiration and criticism as he continually rebels against traditional ideals of fashion. Yamamoto was a graduate of the prestigious Keio University and Bunka Fashion College in Tokyo, and after two years of designing womenswear, set up his own company, Y's, in 1972. Although his premiere collection was revealed in Tokyo in 1977, Yamamoto only achieved widespread recognition after unveiling his Spring/Summer collection in Paris six years later. Since then, Yamamoto has attained numerous awards, such as the Mainichi Fashion Grand Prix and the Fashion Group of America's Designer of the Year; he is the sole Japanese designer to be awarded the French Ordre des Arts et des Lettres. His name stands confidently alongside the likes of Rei Kawakubo and Issey Miyake, and his ready-to-wear brand has taken its place as functional daywear, stretching beyond the boundaries of fashion's ever-changing trends. In 2002 Yamamoto published *Talking to Myself*, a collection of words and images reflecting his life as an international designer. He currently operates as Takeshi Kitano's costume designer for films such as *Dolls*, 2002 and *Zatoichi*, 2003.

basso&**brooke**

Chris Brooke, born 1974, Newark, Nottinghamshire, UK, and Bruno Basso, born 1978, Santos, Sao Paulo, Brazil, both live and work in London.

Anglo-Brazilian design duo Basso & Brooke shot to fame as a result of their win at the Fashion Fringe Awards, created by Colin McDowell in 2004. Based on their heavily patterned pieces on a variety of fabrics, from chiffon and supple leather to silk satin and *crêpe-de-chine*, their win ensured their successful launch onto the London fashion scene. The label combines the talent of English-born Christopher Brooke with Brazilian designer Bruno Basso. Brooke studied for his BA in Fashion at Kingston University, followed by an MA from Central St Martins, while Bruno Basso studied Advertising at the University of St Cecilia, San Paolo, Brazil. The designers are commended for the imaginative use of cutting and tailoring, creating bespoke garments for every uniquely engineered print. Aside from their seasonal collections, shown worldwide, they run a separate cruise line. Backed by Italian fashion group Aeffe, the collection is currently stocked in 55 stores worldwide. Basso & Brooke are celebrating their second New Generation Sponsorship nomination in the Best Newcomer category at the British Fashion Awards.

eley**kishimoto**

Mark Eley, born 1968, Bridgend, Wales, and Wakako Kishimoto, born 1965, Sapporo, Japan, both live and work in London.

Eley Kishimoto was founded in 1992 by Wakako Kishimoto and Mark Eley. Kishimoto earned her BA in Fashion and Print from Central St Martins in 1992, whilst Eley graduated from Brighton Polytechnic in 1990 with a BA in Fashion and Weave. The pair earned international acclaim for designing prints for Joe Casely-Hayford, Hussein Chalayan and Alexander McQueen, amongst many others. In 1995, Eley Kishimoto produced their first collection, *Rainwear*, a range of waterproof coats, umbrellas and gloves made from PVC coated fabrics. Their first on-schedule show was presented in Autumn/Winter 2001, and included a series of afternoon tea parties, rather than a traditional catwalk presentation. The label has also included crockery, furniture, wallpaper, luggage, lingerie and sunglasses. Eley Kishimoto are commissioned by international designers such as Mark Jacobs and Louis Vuitton, but despite the scale of their industry, the couple work quietly with a crew of craftspeople from a workshop in Brixton, South London.

hamish**morrow**

Born 1968, South Africa, lives and works in London.

Hamish Morrow came to London in 1989 to study at Central St Martins, but left the course to begin a career designing and pattern-cutting. In 1996, he returned to his studies, completing an MA in Menswear Fashion Design at the Royal College of Art. After graduation, Morrow began working at Byblos in Milan, under the creative direction of John Bartlett. He returned to London to launch his own label, catching the attention of the press with his second collection, for Autumn/Winter 2001. In October 2001, Hamish was invited to New York by Henri Bendel, who created an installation of his Spring/Summer 2002 collection in-store, and devoted their Fifth Avenue window entirely to his designs, as part of a celebration of London Fashion Week. Morrow is generally regarded to be an avant-garde experimentalist and fashion 'thinker', and thrives on inventive collaborative projects in music, visual arts and digital media. His designs take much of their inspiration from these evolving media, fused with elements of sportswear and extreme sport. He has also previously collaborated with a number of internationally renowned design houses, including Haute Couture at Louis Feraud, Menswear at Fendi and Mariuccia Mandelli at Krizia.

pucci

Born 1914, Naples, Italy, died 1992.

Known as the "Prince of Prints", Emilio Pucci was born to an aristocratic family as Marchese di Barsento a Cavallo, and began his career in fashion in his 30s. After spending his younger life as an athlete and a scholar, with a PhD in Political Science, it was a chance photograph of him wearing ski clothes of his own design in *Harper's Bazaar* that launched his career in ready-to-wear. Opening his own couture house in 1950 after the magazine asked him to design some winter clothes for women, his designs soon garnered worldwide acclaim for their bold, geometric prints and dazzling use of colour. He is credited for introducing wrinkle-free printed silk dresses and 'Capri pants' to the world. His signature patterns reached outer space when the Apollo 15 brought a flag he designed to the moon. His daughter Laudomia Pucci continued to design under the Pucci name following his death in 1992. The Pucci brand was acquired by the French firm Louis Vuitton Moet-Hennessy Group (LVMH) in 2000 and has been associated with designers that include Julio Espada and Christian Lacroix. In 2005, British designer Matthew Williamson was named artistic director, while Laudomia Pucci continues as image director; under their leadership, the Pucci name continues to be synonymous with remarkable style and signature prints.

jonathan**saunders**

Born 1977, Glasgow, lives and works in London.

Fashion print designer Jonathan Saunders graduated from the Glasgow School of Art in 1999 with a BA in Printed Textiles, before going onto receive his MA in Printed Textiles from Central St Martins in 2002. He is known for his original use of print design, shape and unusual use of the colour spectrum, which won him the Lancôme Colour Award in 2002. Each garment carries its own unique print design, and draws upon traditional silk screening techniques. One of London's most promising young designers, his 2003 debut collection attracted attention for its kaleidoscopic forms and innovative use of screen printing. He has provided consulting for some of the biggest fashion houses in Europe, is a regular fixture at London Fashion Week, and continues to expand his range, with the unveiling of his swimwear line in 2005.

zandra**rhodes**

Born 1940, Chatham, Kent, UK, lives and works in London.

Zandra Rhodes was introduced to the world of fashion by her mother, a fitter for the Paris fashion house of Worth and later a lecturer at Medway College of Art. Rhodes was formally educated in Print and Textile Design at The Royal College of Art in London. In 1969 she set up her own retail outlet and took her collection to New York, where she was featured in American *Vogue*. She then started selling to shops such as Henri Bendel and Neiman Marcus, and was given her own area in Fortnum and Mason, London. The recipient of numerous awards, Rhodes was made a Commander of the British Empire in 1997 in recognition of her contribution to fashion and textiles, and has been awarded eight honorary doctorates, from the University of Westminster and London Metropolitan University, among others. Since 1998 she has created an exclusive collection for Liberty of London and now also sells from her exclusive salon in Bermondsey Street, above the Fashion and Textile Museum, which she founded. She sells to other fashionable stores in London including Harrods and Selfridges, and all over Europe in locations from Milan to Madrid to Moscow. Recently she designed the costumes for Mozart's *The Magic Flute* and the sets and costumes for *The Pearl Fishers* for San Diego Opera in her own inimitable style. A retrospective of Rhodes' work, with an accompanying catalogue, *Zandra Rhodes: A Lifelong Love Affair With Textiles*, opened at the Fashion and Textile Museum in 2006.

renato**termenini**

Born 1972, Cremona, Italy, lives and works in London.

Arriving in London in 1996 with ambitions to secure a place within the hallowed ranks of British fashion and design, Renato Termenini left his native Italy, where he had studied tailoring and garment technology, to undertake a BA in Fashion Design at Central St Martins in 2000. With his inventive approach to fashion, and adept design and printmaking skills, Termenini not only made a lasting impact on his tutors, but won the Levi's Bursary Design Award in 2002. By the time he had graduated with Honours in Fashion Design and Print, Termenini had already collaborated with celebrated British designers Robert Cary-Williams and Alexander McQueen. Upon graduating, a collaboration with Pringle of Scotland ensued, working on their womenswear collection. Termenini launched his eponymous womenswear label in late 2004, attaining media attention for his debut collection, and acclaim from the London Fashion Forum who selected him to present at their 2005 event, *Profile*.

carole**waller**

Born 1956, lives and works in Bath, UK.

Having earned a BA Honours in Painting at Canterbury College, and an MA from Cranbrook Academy of Art, Michigan, USA, Carole Waller describes her printed collections as wearable 3-D art. With a strong eye for colour and use of bold hand-drawing, she transform her canvases into garments, using various techniques from screen printing, collage and appliqué to create a seasonal range of unique, classically cut clothing. Having exhibited internationally since 1987, she has attended Fashion Week in both Paris and New York from 1990–2003, and sold her collections in boutiques worldwide. Her work can be viewed in the Costume and Textile Collection at the Victoria and Albert Museum, the Crafts Study Centre Collection, Surrey, and in Birmingham City Art Gallery. Also represented by Contemporary Applied Arts, London and Julie Gallery, New York, she has received awards from the Crafts Council, the BKCEC (British Knitting & Clothing Export Council), and the Michigan Council for the Arts.

shoto**banerji**

Born 1958, Kolkata, India, lives and works in London.

After working for several years as a nursery school teacher, Shoto Banerji moved from India to London in 1996, where she was accepted by Central St Martins College of Art and Design. Graduating three years later with a BA in Textile Design, she returned to India to join the Rehwa Society as a designer. Here, she worked alongside the handloom weavers of Central India, blending traditional skills with new materials and textiles, developing products better suited to an international market. During her time in India, Banerji also became a guest faculty member of the Mumbai National Institute of Fashion Technology, working with the graduating class for the post-graduate programme in Knit. In 2001, Banerji travelled to Belgium where she lived and worked for a number of years. Here, still influenced by the rural weavers of India, she began to develop her own collections, exhibiting at the Galerie de la Madeleine in Brussels in 2002. The collection, inspired by the colouring of Indian birds, was later presented as a number of solo shows in London at the Image d'Or and Cecilia Colman Galleries, in Antwerp at the H-Art Gallery, and in Brussels at Galerie 21. Presently, Banerji is designing her next collection, innovatively bringing together the work of the Indian handloom weavers with new yarns such as Thai silk, Belgian linen and Italian wool, to be previewed towards the end of 2006.

liz**collins**

Born 1968, Washington DC, lives and works in Providence, Rhode Island.

Liz Collins is an artist and designer, recognised internationally for her use of machine knitting to create ground-breaking clothing, textiles, and 3-D installations. After four years as an independent designer of seasonal ready-to-wear collections in New York City, Collins returned to her alma mater, Rhode Island School of Design, as an Assistant Professor in the Textile Department. Collins currently designs knitwear under her own label, which she sells at trunk sales and select boutiques in New York and Tokyo. She has also collaborated with other designers, producing signature knit pieces and collections for them. A member of the Council of Fashion Designers of America, Collins was featured last winter in the CFDA's inaugural designer showcase with the QVC television network, and was recently cited by *The New York Times* as a "designer with many industry accolades".

márcia**ganem**

Márcia Ganem began her career as a designer in Salvador, Bahia, combining synthetic materials, such as nets made from polyamide fibres, with organic materials, such as wood, semi-precious stones and natural beads. By combining technology and craft, Ganem is able to create unique, modern and quasi-artistic pieces. Her main influence is the rich popular culture of Bahia and its pervasive African influence: especially the mysteries of the Candomblé rituals, from which she finds inspiration in the xequeré, a percussion instrument covered with small beads, and in the characters and ornaments of the orixás, the Candomblé gods. Ganem has a studio and a shop in Pelourinho, in the old part of Salvador, and has acquired a good name in the Brazilian fashion world because she is not a 'slave' to international trends. Her clothes and jewellery have been shown in the fashion sections of British papers such as *The Sunday Times* and *The Daily Telegraph*. Márcia Ganem continues to present her collections at the Rio Fashion Week.

louise**goldin**

Lives and works in London.

Young designer Louise Goldin has launched an active design career since receiving her MA in Fashion and Knitwear at Central St Martins in February 2005. Her knitwear label, building upon the success of her MA show, has sold exclusively in Selfridges in London, and has been exhibited at Indigo in Paris in September 2005 through a prize sponsored by Textprint. Also in 2005, she received the Deutsche Bank Pyramid Award for her promise in the worlds of fashion and business, and in 2006 she was selected as part of the *New Generation* exhibition at London Fashion Week, during which her works were exhibited at the Natural History Museum. Goldin's already considerable design capacities are sure to expand as she continues to establish herself in fashion and textile design.

shirin**guild**

Born 1946, Tehran, Iran, lives and works in London.

Shirin Guild's Iranian menswear inspired clothing line engages both her personal and ethnic history, as well as her appreciation of volume and texture. Although she was tutored at Central St Martins for six months, Guild's design and craftwork—often characterised by their loose-fit, layers, and boxy simplicity—are largely self-taught. After years of making and designing her own clothing, Guild and her husband Robin launched the Shirin Guild label in London in 1991. By directly sourcing all of her garments' distinct yarns and fibres—from papers to silk to linen (from Italy, Japan and Iran)—Guild maintains an uncommon standard of quality in her creations. Her designs and carefully selected materials came to define a genre of 'ethnic minimalism' in British fashion in the 1990s, and her work is on permanent display at the Victoria and Albert Museum in London, the Brighton Museum and Art Gallery, and the Fashion Institute of Technology in New York.

kanako**kajihara**

Born 1973, Sapporo, Japan, lives and works in Hokkaido, North Japan.

Kanako Kajihara has a BA in Textile Design from Tokyo's Tama Art University, and an MA in Constructed Textiles from the Royal College of Art, London. Completed in 1998 and 2003, respectively, these qualifications have set the young designer up for a prosperous and already fruitful career in textiles, enabling her to work alongside some of the fashion industry's greatest names. Following her initial degree, Kajihara spent time working with world renowned designer Issey Miyake, where she was given the responsibility of developing printed textiles and pleat techniques. Hoping to independently study and create a number of more innovative fabrics, after three years with Miyake, she went on to win the Texprint 2005 Breaking New Ground Prize, for work developed on her MA course. Interested in natural concepts of 'adaptation' and 'attraction', Kajihara based her latest collection upon the idea of nature's surfaces and the ways in which they adapt to their surrounding environment. Inspired by the richness and opulence of the Guatemalan jungles, she has created an impressive body of work that continues to attract attention from textile designers worldwide. Having worked with numerous companies including Union Design Inc, Tokyo, while continuing to push the boundaries of conventional textile design, Kajihara is currently preparing for an exhibition in London entitled *Re-Series*.

lainey**keogh**

Born 1957, Dublin, where she continues to live and work.

A combination of rural craft and urban technology lies at the heart of Lainey Keogh's knitwear design, which is hand-made around Ireland by women in their own homes. Keogh herself started out hand-knitting sweaters for friends before deciding to devote herself full-time to fashion in the mid-1980s. In 1989, she was presented with the Prix de Coeur by Christian Lacroix in Monte Carlo for her work with Irish linen. She worked as costume designer for John Boorman's 1995 film *Two Nudes Bathing*, for which she won a Cable Ace Award. The designer received critical acclaim in *Vogue* for the first showing of her self-titled label at London Fashion Week in February 1997. Keogh's lines are stocked in discerning stores such as Browns in London, Biffi in Milan, Ron Herman of Los Angeles, Bergdorf Goodman in New York and Brown Thomas in her native Dublin.

jurgen**lehl**

Born 1944, Poland, lives and works in Tokyo.

Polish-born and French-schooled Lehl entered the world of textile design while living and working freelance in France in the 1960s. In 1969, he moved to New York, but soon after discovered what was to become his permanent home in Japan, where he moved in 1971. He founded Jurgen Lehl Co. in 1972 and showcased his first ready-to-wear collection in 1974. Lehl continues to create unique men's and women's clothing, accessories, and jewellery, and is a regular participant in many solo and group exhibitions around the world. His exhibitors include the Cooper-Hewitt Museum in New York, the Museum für Angewandte Kunst in Cologne, and the Gemeentemuseum den Haag in The Netherlands, and he has received numerous awards for creative textile and advertising designs. Although his designs have not always translated back to the West, he continues to showcase and sell his earth-inspired collections throughout Japan and East Asia.

missoni

Ottavio Missoni, born 1953, Dalmatia, Italy, Rosita Jelmini Missoni, born 1931, Lombardy, Italy, Angela Missoni, born 1958, Milan, and Luca Missoni, born 1956, Milan, where all live and work.

Founded in 1953 in Gallarate in Italy, Missoni has remained a family-run business throughout its 40 year history. Husband and wife team Ottavio and Rosita first produced knitwear for the Rinascente boutique in 1954, and introduced the Missoni label in 1958. The technical innovation of those early days, experimenting with knitting machines and techniques to produce their abstract patterning in appealing colour combinations, established the foundation of a signature style that made the Missoni brand world famous. Simple, elegant cuts were realised in a rich range of hues, interwoven with metallic threads and lurex. The Missoni SpA workshop and factory was established in 1968, and the first boutiques opened in 1976 in Milan and New York. Throughout the 1960s and 70s, the label's layered, mis-matching, casual pieces embodied the Italian chic so fitting to the post war lifestyle. The couple's daughter, Angela, introduced her first collection in 1992, and took over the womenswear lines in 1997. Since taking the reins, she has introduced a footwear line, in 1998, and in 1999, M Missoni was launched as a more affordable line of the company's luxury brand. With son Luca Missoni designing the menswear collections, Rosita and Ottavio's children are ensuring that the family name will retain its place at the forefront of Italian fashion.

anne**maj**nafar

Born 1965, Roskilde, Denmark, lives and works in Als, Denmark.

Anne Maj Nafar first studied textiles at the Hellerup Textile College in Copenhagen, Denmark. Graduating in 1991 with a BA in Textiles and Teaching, she went on to study Industrial Design at TEKO for which she was awarded a BA in 1996. Impressively, Nafar also received a Diploma in Technology in Knitwear from the Stoll Company Course Centre in Reutlingen, Germany in 1992 and, more recently, in 2006, an MA in Fashion Design Technology from the London College of Fashion, University of Arts. Additionally, Nafar has worked with a number of design companies in Denmark including Micha, Belika and Filati. With a penchant for authenticity and retro design, Nafar has sought, throughout her career, to develop ways in which designers can create beautiful yet 'imperfect' clothing, without the use of mass-production and intensive technology. Continuing her pursuits in this area of interest, Nafar has, since 2001, been a lecturer and Fashion Design Technologist at the International Academy of South Denmark. Here, she continues to learn and inspire.

hikaru**noguchi**

Lives and works in London.

With a degree in graphic design from Musashino Art University and a background in textiles, Tokyo native Hikaru Noguchi has been reworking knitwear in her signature, eclectic style for the past decade. The depth of the furnishing fabric culture in Europe drew Noguchi to design her first collection, for her degree show, in textiles for furniture. From thereon, she progressed to an interest in designing knitted garments. During her studies at Middlesex University she decided to pursue fashion specifically, and forged influential relationships with British designers like Tom Dixon, who offered praise and encouragement. After receiving a grant from the Crafts Council to set up her own business, she showed her scarf and furnishing collection at Chelsea Craft Fair for the first time in 1994 and earned high profile clientele such as Barneys New York. Showing her collection at London Fashion Week in 1997 and in Paris in 1999, she wove her particular brand of knitwear classics into the consciousness of fashion editors and has been featured in magazines like *Vogue*, *Elle*, *Marie Claire*, *Glamour*, *In Style* and others. Elegant and whimsical, Noguchi's designs are sold in boutiques and department stores in London, Paris, New York, Tokyo, Hong Kong and Australia.

jessica**ogden**

Born 1970, Kingston, Jamaica, lives and works in London.

Having been described as the pioneer of salvage fashion, Jessica Ogden's craft orientated look combines a subtle girlishness with a feminine elegance. Brought up in Jamaica by her English parents, Jessica Ogden studied at The Rhode Island School of Design in Providence and The Byam Shaw School of Art, London. Having started her career as a volunteer for Oxfam's NoLoGo project in 1992, reworking old charity shop finds into original garments, she then launched her own line in 1993. Her approach involves reviving antique or distressed fabrics and garments, using hand-patching, quilting, and embroidery, as well as treatments involving paint, needles, cheese graters, hammers and nail guns to create her one-off pieces. While she takes part in London Fashion Week, she prefers to adopt a more unconventional way in which to display her work. Collaborating with photographers and stylists, she creates mixed-media installation pieces to show alongside her collections. She is currently working with the French label APC for her own JO custom line and maintains her strong customer base by participating in various travelling exhibitions.

clare**tough**

Born 1980, lives and works in London.

Having been awarded a first class degree from Chelsea College of Art in 2001, knitwear designer Clare Tough was quickly pounced upon by the fashion press in 2004 with her MA collection at Central St Martins, which was bought by Browns, London. Selected as one of fashion's brightest rising stars, she won Topshop New Generation Sponsorship for 2005. Tough's signature style is a combination of machine- and hand-knits and crochet, in a patchwork mixture of yarns and contrasting colours; she has also worked with leather, suede and rubber in a similar style. With short, tight garments and daring cut-outs she has earned a reputation as a creator of sexy, glamorous pieces. She launched her own label at London Fashion Week in 2006, which is now sold in Journal Standard and Side by Side in Tokyo, Stig-p in Denmark, Espace Mirage in Monaco, Shine in Hong Kong and Banner in Milan, as well as Browns in London.

notes

the hidden art of fashion

1. Many current textile graduates emanate from the same UK establishments, such as the Central School of Arts and Crafts in London, which have evolved into contemporary university departments. These graduates work in the worldwide textiles and fashion industries.

2. The Utility scheme set up by the British government to produce templates for textiles and clothing production in wartime ironically created excellent conditions for good design, and enlisted the skills of many eminent people.

3. See also the work of Lucy Orta, to be showcased in a forthcoming Black Dog Publishing publication in 2007.

4. A cynical view is that its permeation through the price spectrum of fashion has something to do with the ease with which these effects could be emulated in low cost production—unfinished equates to fewer production processes.

5. *Structure and Surface: Contemporary Japanese Textiles*, Museum of Modern Art, New York 1998. *2121: The Textile Vision of Reiko Sudo and Nuno*, James Hockey Gallery, Farnham, Surrey and touring UK 2005.

6. Author's interview with Reiko Sudo, Tokyo, 4 November 2005.

7. Martin Margiela, however, a constant fashion maverick, deconstructs the very notion of a fashion show, and influentially stages his seasonal presentations in derelict and transitory spaces around the environs of Paris. On several occasions Margiela has dispensed with models and used mannequins or photographs, or models just mingled with the people in the streets.

8. Many international designers have also designed ranges for other houses—Martin Margiela for Hermes, Hussein Chalayan for Tse Cashmere and currently Marc Jacobs for Louis Vuitton, Nicholas Ghesqiere for Balenciaga, Antonio Marras for Kenzo. In recent years, more and more liaisons between named fashion designers and British high street brands, such as Julien Macdonald for Marks & Spencer, Stella McCartney for H&M, are now commonplace, as designer wear becomes an achievable aspiration for more people.

9. See "Ritu Kumar-Designing for Queens" in *The South Asian*, January 2001. www.the-south-asian.com/Jan2001/RituKumar

embellished
stephanieaman
1. All quotes are from the author's interview with the designer, March 2006.

ashishgupta
1. All quotes are from the author's interview with the designer, March 2006.

yoshikihishinuma
1. McCarty, Cara and Matilda McQuaid, *Structure and Surface: Contemporary Japanese Textiles*, New York: Museum of Modern Art, 1998, p. 12.

2. McCarty, *Structure and Surface: Contemporary Japanese Textiles*, p. 25, p. 72.

3. Braddock, Sarah E and Marie O'Mahoney, *Techno Textiles: Revolutionary Fabrics for Fashion and Design*, London: Thames and Hudson, 1999, pp. 124–125.

4. Webb, Martin, "Fashion: Brave New World", www.metropolis.japantoday.com

christianlacroix
1. Lacroix, Christian, *Pieces of A Pattern: Lacroix*, London: Thames & Hudson, 1992, p. 32.

2. Weisman, Katherine, "Couture's Special Effects", *Women's Wear Daily*, 7 November 1997.

threeasfour
1. Yaeger, Lynn, "Beyond the Fringe", *The Village Voice*, 23–29 January 2002.

sculpted
sangheechun
1. From the author's interview with the designer, March 2006.

reikawakubo
1. Morris, Bernadine, "From Japan, New Faces, New Shapes", *The New York Times*, 14 December 1982.

2. Sudjic, Deyan, *Rei Kawakubo and Comme des Garçons*, New York: Rizzoli, 1990, p. 80.

junkokoshino
Koshino, Junko, "Junko Koshino's Classical Modernism", Tokyo: Junko Koshino Press Office.

ezinmambonu
1. All quotes are from the author's interview with the designer, April 2006.

rowanmersh
1. From the author's interview with the designer, April 2006.

yohjiyamamoto
1. http://www.japan-zone.com/modern/

2. Yohji Yamamoto as quoted in *The Fashion Book*, London: Phaidon, 1998.

imprinted
hamishmorrow
1. All references are from an interview with Hamish Morrow by Penny Martin, 23 September 2003. http://www.showstudio.com/projects/bot/bot_interview.html

zandrarhodes
1. Polan, Brenda, "The Art of Textiles", *Zandra Rhodes: A Lifelong Love Affair with Textiles*, London: Antique Collectors' Club, 31 July 2005. p. 16.

2. Rhodes, Zandra, *The Art of Zandra Rhodes*, London: Trans-Atlantic Publications, 1995, p. 56.

constructed
lizcollins
1. From the author's interview with the designer, March 2006.

laineykeogh
1. Abadani, Kaveh, "Animal Magic", *Vogue*, 22 February 1999.

jurgenlehl
1. Quotation from Jurgen Lehl in *Fashions* by Jurgen Lehl, The Hague: Gemeentemuseum, 2000, p. 39.

annemajnafar
1. From the author's interview with the designer, March 2006.

hikarunoguchi
1. From the author's interview with the designer, March 2006.

jessicaogden
1. Hirayama, Hakuho, *Sumi-E*, Tokyo: Kadansha International Ltd., 1979, p. 7.

claretough
1. From the author's interview with the designer, March 2006.

acknowledgments

Fashioning Fabrics would not have been possible without the research, writing, design and dedication of numerous people. The passion and commitment of Sandy Black and Elyssa Da Cruz, was the driving force behind this publication, as was the stunning imagery so graciously provided by the numerous designers and their representatives. Anna Chapman and Shoto Banerji put together the initial list of designers to be included, with Shoto Banerji's hard work continuing on with picture research and administration that was indispensable to the completion of the publication. Thanks also go to the numerous writers who contributed text including Natalie Bell, Beccy Clarke, Indigo Clarke, Laura McLean-Ferris, Sabrina O'Cock, Jennifer Trak, and Zuki Turner. I am also grateful to Sion Parkinson for travelling around London, taking photographs within several designers' studios, giving the book a unique glimpse into the creative process. Amy Sackville is due thanks for her attention to detail in the final proofing of the book. Lastly, the designer Emilia Gómez López and her assistant Lisa Drake have spent many hours dealing with the vast amount of visual material included in the book to create a beautiful design that allows the variety of work to maintain its individuality.

Oriana Fox

Related titles from Black Dog Publishing:

Unclasped Contemporary British Jewellery	1 901033 35 X
New Directions in Jewellery	1 904772 19 6
New Directions in Jewellery II	1 904772 55 2
The Cutting Edge of Wallpaper	1 904772 56 0
Making Stuff An Alternative Craft Book	1 904772 61 7

Black Dog Publishing
Architecture Art Design Fashion History
Photography Theory and Things

www.bdpworld.com

Edited by Sandy Black
Initial Research by Anna Chapman and Catherine Heygate
Picture Research by Shoto Banerji
Production by Oriana Fox @ BDP
Designed by Emilia Gómez López @ BDP with the assistance of Lisa Drake

Black Dog Publishing Limited
Unit 4.4 Tea Building
56 Shoreditch High Street
London
E1 6JJ

Tel: +44 (0)20 7613 1922
Fax: +44 (0)20 7613 1944
Email: info@bdp.demon.co.uk
www.bdpworld.com

All images by Anthea Simms are courtesy the photographer. All other images are courtesy the
designer unless otherwise stated.

Front cover image: Rowan Mersh, prototype fabric 2005.
Back cover image: Christian Lacroix, Autumn/Winter 2001, photo: Gilles Kervella.